Cambridge Elements

Elements in Decision Theory and Philosophy
edited by
Martin Peterson
Texas A&M University

BARGAINING THEORY

Peter Vanderschraaf
University of Arizona

CAMBRIDGE
UNIVERSITY PRESS

CAMBRIDGE
UNIVERSITY PRESS

Shaftesbury Road, Cambridge CB2 8EA, United Kingdom

One Liberty Plaza, 20th Floor, New York, NY 10006, USA

477 Williamstown Road, Port Melbourne, VIC 3207, Australia

314–321, 3rd Floor, Plot 3, Splendor Forum, Jasola District Centre, New Delhi – 110025, India

103 Penang Road, #05–06/07, Visioncrest Commercial, Singapore 238467

Cambridge University Press is part of Cambridge University Press & Assessment, a department of the University of Cambridge.

We share the University's mission to contribute to society through the pursuit of education, learning and research at the highest international levels of excellence.

www.cambridge.org
Information on this title: www.cambridge.org/9781108706681

DOI: 10.1017/9781108588638

First published 2023

A catalogue record for this publication is available from the British Library.

ISBN 978-1-108-70668-1 Paperback
ISSN 2517-4827 (online)
ISSN 2517-4819 (print)

Bargaining Theory

Elements in Decision Theory and Philosophy

DOI: 10.1017/9781108588638
First published online: February 2023

Peter Vanderschraaf
University of Arizona

Author for correspondence: Peter Vanderschraaf,
pvanderschraaf@email.arizona.edu

Abstract: The Nash bargaining problem provides a framework for analyzing problems where parties have imperfectly aligned interests. This Element reviews the parts of bargaining theory most important in philosophical applications, and to social contract theory in particular. It discusses rational choice analyses of bargaining problems that focus on axiomatic analysis, according to which a solution of a given bargaining problem satisfies certain formal criteria, and strategic bargaining, according to which a solution results from the moves of ideally rational and knowledgeable claimants. Next, it discusses the conventionalist analyses of bargaining problems that focus on how members of a society can settle into bargaining conventions via learning and focal points. In the concluding section this Element discusses how philosophers use bargaining theory to analyze the social contract.

Keywords: bargaining problem, social contract, axiomatic solutions, strategic solutions, bargaining conventions

ISBNs: 9781108706681 (PB), 9781108588638 (OC)
ISSNs: 2517-4827 (online), 2517-4819 (print)

Contents

1 Introduction

Bargaining theory is a part of mathematical social science that explores how agents whose interests can be somewhat misaligned might reach some mutually beneficial arrangement. Bargaining theory is distinctive in that a given *Nash bargaining problem* is fully characterized by the combinations of individual utilities such agents can achieve by interacting. John Nash and Howard Raiffa laid the foundations of bargaining theory in the early 1950s, Nash in a pair of now classic journal articles (1950, 1953) and Raiffa in his doctoral thesis (1951), which Raiffa later published in revised form (1953). One major part of this theory, now known as *axiomatic bargaining theory*, considers which outcomes of a bargaining problem satisfy certain desirable formal properties. Axiomatic bargaining theory is part of *cooperative game theory*, the branch of game theory that analyzes how utilities are distributed among sets of agents constrained by binding agreements. Another major part of bargaining theory, now known as *strategic bargaining theory*, considers which outcomes of a bargaining problem rational agents equipped with sufficient mutual knowledge of their situation might reach via their interactions. One can view strategic bargaining theory as a part of classical *noncooperative game theory*, the branch of game theory that analyzes the consequences of the interactions of strategies chosen by rational agents unconstrained by binding agreements. Only a few years following Nash's and Raiffa's foundational works, Thomas Schelling (1960) considered an alternate approach to analyzing bargaining problems in terms of *learning* and *focal points*. Schelling's work established a part of bargaining theory one might call *conventionalist bargaining theory*, since this part of the theory examines how agents with somewhat limited knowledge who engage in a bargaining problem might reach one of many available *bargaining conventions*. One can view conventionalist bargaining theory as a part of *evolutionary game theory*, the branch of game theory that examines how populations can settle into stable patterns of behaviors that interact.

Since Nash and Raiffa defined bargaining problems entirely in terms of utility spaces, bargaining theory is a remarkably flexible framework for analyzing problems where agents can gain by arriving at some sort of agreement. Nash shows by example that one can use bargaining theory to select a final equilibrium outcome of an exchange economy. Richard Braithwaite was the first professional philosopher to make use of bargaining theory, and Braithwaite used this theory to analyze a specific *fair division problem* in his Cambridge lecture *Theory of Games as a Tool for the Moral Philosopher*.[1] The specific rule

[1] Braithwaite's 1954 lecture, which he subsequently published ((1955) 1994), was his inaugural lecture upon being appointed Knightbridge Professor of Moral Philosophy at the University of Cambridge.

for fair division Braithwaite defended in his Cambridge lecture is a very simple example of a social contract. A bargaining problem models in terms of utilities the situation of agents who are in the *circumstances of justice*. Agents in the circumstances of justice are generally capable both of affecting each other's prospects by their actions and of cooperating for mutual benefit with respect to some baseline state of affairs, and they are likely to have conflicting preferences over alternative cooperation schemes.[2] The rules that implement the terms of a cooperation scheme constitute a social contract for such agents. In the years following Braithwaite's lecture, some philosophers and social scientists have followed Braithwaite's lead and employed bargaining theory for developing parts of more complex social contracts.

In this Element I review the parts of bargaining theory most important in philosophical applications, and to social contract theory in particular. When I introduce any one of these parts of bargaining theory, I review any necessary accompanying specific mathematical vocabulary or formalisms. So this Element is self-contained and should be accessible to readers familiar with elementary set theory and algebra but who do not necessarily have a previous acquaintance with bargaining theory or game theory. I proceed as follows: In Section 2 I present four examples that are problems where agents can cooperate to their mutual advantage but may have difficulty agreeing upon just how they might cooperate. I present these examples in an informal manner. These four examples help to motivate the more formal discussions of the Nash bargaining problem in the later sections. In Section 3 I review how a bargaining problem is characterized as a *feasible set* of points together with a *nonagreement point*, all of which are in utility space. Here I present specific examples of bargaining problems based upon the Section 2 examples. I also discuss some structural properties of certain classes of bargaining problems, including especially *comprehensiveness* properties. Furthermore, I outline how one can construct a bargaining problem from a noncooperative *basis game*, and how conversely one can construct a non-cooperative *demand game* from a bargaining problem. In Section 4 I discuss *solution concepts*. Bargaining theorists have proposed a great many different solution concepts for bargaining problems. The solution concepts I discuss in this section are those I believe are most important for bargaining theorists

[2] Rawls introduces the term "circumstances of justice" in *A Theory of Justice* (1971, 126). Rawls reviews these circumstances and credits Hume with presenting an especially clear statement of them (1971, 126–130). Hume gives his most detailed presentation of the circumstances of justice in *An Enquiry Concerning the Principles of Morals* 3.1 and also reviews these circumstances in *A Treatise of Human Nature* 3.3.2: 15–18. I discuss the circumstances of justice and propose a game-theoretic summary of these circumstances in Vanderschraaf, 2019, 85–117.

and for social contract theorists. Here I concentrate mainly on definitions and illustrations of these solution concepts. Still, for the *Nash*, the *Egalitarian* and the *Kalai–Smorodinsky* solutions I also show how geometric constructions of these solutions in a specific two-agent bargaining problem based upon Braithwaite's lecture help to motivate these concepts. In the next two sections I turn to approaches to justifying various solutions. In Section 5 I discuss rational choice approaches to justifying solution concepts. Here I review parts of axiomatic and strategic bargaining theory and how these two rational choice theories can complement each other. I show that while a number of solution concepts are characterized by different sets of intuitively appealing axioms, no solution concept satisfies all of a fairly small set of such axioms. I conclude that in the end rational choice bargaining theory does not vindicate a single solution concept for all bargaining problems. Rather, rational choice bargaining theory is valuable mainly as a means for identifying which properties one ultimately believes a class of bargaining problems one is analyzing must satisfy. In Section 6 I discuss parts of the conventionalist analysis of bargaining problems. I argue that this conventionalist analysis complements the rational choice analysis of bargaining problems in interesting ways. I consider how agents who engage repeatedly in a bargaining problem can learn from their experiences in these engagements to follow certain bargaining conventions. I discuss how some studies of inductive learning, including a study of my own, show how for certain classes of bargaining problems some bargaining conventions that approximate certain well-known axiomatic solutions tend to emerge more frequently than others. What agents learn inductively can lead them to regard the bargaining conventions that match these axiomatic solutions as *focal points* or *salient* conventions. In the concluding Section 7 I discuss the main philosophical application of bargaining theory – namely, in the analysis of the social contract. Here I draw an overall conclusion analogous to the one I draw regarding rational choice bargaining theory in Section 5. I argue that the bargaining problem has special advantages as a tool for analyzing social contracts. I note that the nonagreement point of a bargaining problem summarizes the plight of agents who remain in a State of Nature and that alternative points of the feasible set summarize how these agents can all benefit at different alternative social contracts. But I also argue that no one solution concept is the concept to which all satisfactory social contracts must conform. Rather, different bargaining problem solution concepts can help one to identify which properties one will require of a given social contract for a given social situation.

2 Motivating Problems

Chocolate Cake

Claudia, a classical cellist, and Laura, a classical violinist, are presented with a chocolate cake prepared by a local baker immediately following one of their performances together. Laura and Claudia have to decide how they will dispose of this cake. Suppose each can claim any share of the cake. Their claims are *compatible* if there is enough cake to provide each her claimed share. If Claudia and Laura issue compatible claims, then each indeed takes her claimed share. But if their claims are incompatible, they fight and the cake spoils in the meantime.

This Chocolate Cake Problem has a *Nash demand game* structure. In a Nash demand game, agents issue their claims, and each receives her claim if their joint claims are compatible and otherwise must follow her part of a *nonagreement outcome*. The baker's cake is a homogeneous good that presumably can be divided into arbitrarily small pieces. So Claudia (Agent 1) and Laura (Agent 2) each have infinitely many *pure strategies* in their Nash demand game, each of which is to claim some share $x_i \in [0, 1]$. At the nonagreement outcome neither Claudia nor Laura receives any cake and they also fight, so both bear costs of fighting neither one would have to bear were she to simply claim none of the cake.

On the surface, solving the Chocolate Cake Problem seems simple enough: The problem is solved when each musician claims half. If one assumes that the utility for each increases in the amount of cake she receives, then to claim half is the best response for each against the other's claim of half. If Claudia claims $x_1 = \frac{1}{2}$ and Laura claims $x_2 = \frac{1}{2}$, then each receives half of the cake. If Claudia changes her claim on her own to some value $x_1' < \frac{1}{2}$, then she receives less than half. And if Claudia changes her claim to $x_1'' > \frac{1}{2}$, then she receives no cake at all and also fights with Laura, an outcome even worse for Claudia than some outcome where she claims $x_1 = 0$. Similarly Laura is strictly worse off if she unilaterally changes her claim from $x_2 = \frac{1}{2}$. Since each does best to stick with her part of the claim profile $\boldsymbol{x}_{\frac{1}{2}} = \left(\frac{1}{2}, \frac{1}{2}\right)$, $\boldsymbol{x}_{\frac{1}{2}}$ is a *Nash equilibrium*. Indeed, $\boldsymbol{x}_{\frac{1}{2}}$ is a *strict* Nash equilibrium since each fares strictly worse if she changes her claim on her own.[3] But the very same argument supports Claudia and Laura following each claim profile of the form

[3] For a game with the n agents of the set $N = \{1, \ldots, n\}$ where at a given strategy profile $\boldsymbol{s} = (s_{1i_1}, \ldots, s_{ni_n})$, $u_i(\boldsymbol{s})$ is Agent i's *payoff* at $\boldsymbol{s}$, a strategy profile $\boldsymbol{s}^* = (s_1^*, \ldots, s_n^*)$ is a Nash equilibrium if for each $i \in N$,

(N1) $E_i(u_i(\boldsymbol{s}^*)) \geq E_i(u_i(s_1^*, \ldots, s_{i-1}^*, s_{ij_i}, s_{i+1}^*, \ldots, s_n^*))$

for any strategy s_{ij_i} that Agent i can follow – that is, Agent i's part of $\boldsymbol{s}^*$ is a *best response* for Agent i in terms of expected payoff given that the other agents of N follow their parts of $\boldsymbol{s}^*$. $\boldsymbol{s}^*$ is strict if each of the (N1) inequalities is strict; in other words, Agent i's part of $\boldsymbol{s}^*$ is Agent i's unique best response given that the other agents of N follow their parts of $\boldsymbol{s}^*$.

$x_z = (z, 1 - z)$ for any $z \in [0, 1]$. Laura and Claudia have infinitely many different strict Nash equilibria of the form x_z available to them. Why should one suppose they will follow the $x_{\frac{1}{2}}$ equilibrium rather than some other equilibrium? Indeed, why suppose they will follow any equilibrium at all?

Nash's Traders

Nash illustrates the bargaining problem he introduces with a barter exchange economy example. Two individuals, Bill and Jack, each begin with different sets of toys, any subsets of which they can trade. The utility for each of any set of toys each possesses at end of trading is the sum of the individual utilities for him of each of the individual goods in this set. Table 1 summarizes their individual utilities over the toys available for trading (Nash 1950, 160–161).

Given their initial endowments, Bill's utility is 12 and Jack's utility is 6. Both can do better if they do make a trade. For example, if they simply swap their initial endowments, Bill's utility is then 22 and Jack's utility is then 10. But swapping their initial endowments is only one of many possible trades for Jack and Bill. Which, if any, of these trades should they make in the end?

In Nash's example each of the 512 logically possible deterministic trades defines a payoff vector defined by Bill's and Jack's respective utilities at the

Table 1 Utilities for toys for Nash's traders

Bill's initial endowment	**Bill's utility**	**Jack's utility**
book	2	4
whip	2	2
ball	2	1
bat	2	2
box	4	1
Jack's initial endowment		
pen	10	1
toy	4	1
knife	6	2
hat	2	2

outcome of this trade.[4] Nash concludes that according to his analysis of their bargaining problem, Bill and Jack will make the trade where Bill gives Jack the set {book, whip, ball, bat} and in exchange Jack gives Bill the set {pen, toy, knife}, so that Bill's final utility is 24 and Jack's final utility is 11. In Nash's Trading Problem different agents might start and finish with a variety of distinct goods, none of which happen to be individually divisible.

Braithwaite's Neighboring Musicians

In his Cambridge lecture, Braithwaite presents an example of a division problem that is in some respects more complex than Claudia and Laura's Chocolate Cake Problem. Luke, a classical pianist, and Matthew, a jazz trumpeter, live in neighboring apartments. They have only the same hour each evening free for playing their instruments. Unfortunately for them, the wall their adjacent apartments share does not protect either at all well from hearing what his neighbor does. Each can hear the other's playing almost as well as his own. Each finds just listening to the other play more pleasant than silence. But each would rather play his own instrument undisturbed than listen to his neighbor play undisturbed. And they both would like to avoid the cacophony that results if they play simultaneously. After describing relevant details of Matthew and Luke's situation,[5] Braithwaite puts himself in the place of their arbiter: "Suppose that they put to me the problem: Can any plausible principle be devised stating how they should divide the proportion of days on which both of them play, Luke alone plays, Matthew alone plays, neither play, so as to maintain maximum production of satisfaction compatible with fair distribution?" ((1955) 1994, 9).

Braithwaite proposes to analyze the problem from the perspective of an arbiter rather than from the perspectives of claimants who try to resolve their problem themselves, as they would in a Nash demand game. Braithwaite's example resembles the Chocolate Cake division problem in that two claimants vie over a quantity of a homogeneous and infinitely divisible good, in this case time for playing one's instrument undisturbed. But Braithwaite's division problem has certain important asymmetries not present in the Chocolate Cake Problem that become evident when he presents a mathematically precise formulation. One especially important asymmetry is that the cacophony of simultaneous playing is the worst possible outcome

[4] A trade for Bill and Jack is deterministic when, perhaps after some negotiation, each simply chooses and gives to the other some subset, which might be empty, of his initial endowment and receives some possibly empty set of the other's goods in return. So the null trade where Bill and Jack simply keep their initial endowments, and effective unilateral gifts where Bill (Jack) gives Jack (Bill) some of his original endowment and receives nothing in return are counted as deterministic trades. For some of these 512 different trades, the resulting payoff vectors coincide.

[5] Braithwaite declares that neither Luke nor Matthew can move away from his apartment and that neither can take legal action against the other for making noise in his own apartment ((1955) 1994, 8).

for Luke, but not for Matthew. Given the formal structure of the Braithwaite Problem it is by no means obvious which, if any, of the possible time division schemes are satisfactory resolutions.

Braithwaite was perhaps the first to directly apply Nash's bargaining problem to a problem of dividing some good among a set of claimants. Braithwaite may also have been the first to recognize the more general potential value of applying game theory in social philosophy. Indeed, in his closing remarks Braithwaite speculated that game theory might ultimately transform moral philosophy ((1955) 1994, 54–55). Braithwaite's contributions in his Cambridge lecture were underappreciated at the time and for some years after.[6] In the first part of his 1989 work *Theories of Justice*, Brian Barry presents a remarkable discussion of the Nash bargaining problem, using Braithwaite's neighboring musicians example and Braithwaite's own arguments for a particular proposed solution to this problem as motivation. Barry argues that one can use a Nash bargaining problem to summarize the simplest small-scale case of justice viewed as a system of mutual advantage, where two agents are in dispute over a single issue (1989, 9–12). Barry ultimately rejects the general theory of justice as mutual advantage (1989, 162–163, 373). Nevertheless, Barry's discussion of the bargaining problem remains invaluable, and raised the profile of Braithwaite's original contributions among contemporary social philosophers.

Hume's Thirsty Individuals

In *A Treatise of Human Nature*, David Hume gives a watershed analysis of *convention*. Hume maintains that what is fundamental to a conventional practice for a society is a supporting system of mutual expectations that members of this society follow this practice on account of the mutual benefits they enjoy from their reciprocal compliance.[7] As David Lewis observes in his contemporary classic *Convention: A Philosophical Study*, Hume's analysis closely resembles Lewis' own analysis of a convention as one of a plurality of equilibria in a game that the agents involved follow on account of their *common knowledge* that all expect each other to follow *this* equilibrium, and no other (Lewis 1969, 3–4).[8]

[6] Important exceptions include Luce and Raiffa (1957, 145–150) and Lucas (1959). Rawls (1958, 176–177 n. 12) also gave an early response to Braithwaite's lecture, arguing that the actual moral one should draw from Braithwaite's example and analysis is that game theory is useful for analyzing moral concepts only if one incorporates more than simply the preferences and relative positions of the agents into the analysis. Rawls acknowledges that Braithwaite does this via the procedure by which he selects his solution. As I will discuss below in Section 7, a number of the best-known axioms of bargaining theory formalize general moral concepts.

[7] Hume gives his analysis of convention primarily in *Treatise* 3.2.2–3. Hume also gives a summary of his *Treatise* analysis in *Enquiry* Appendix III.

[8] Lewis gives an analysis of common knowledge in Lewis 1969, 52–57, and his final definition of convention in Lewis 1969, 78–79. Lewis was one of several scholars who independently proposed analyses of common knowledge in the 1960s and 1970s. Informally, Lewis-common

As part of his overall analysis, Hume presents a variety of examples to illustrate how conventions can originate. In one of these examples, three thirsty individuals fall into a dispute.

> Suppose a *German*, a *Frenchman*, and a *Spaniard* to come into a room, where there are plac'd upon the table three bottles of wine, *Rhenish*, *Burgundy* and *Port*; and suppose they shou'd fall a quarrelling about the division of them; a person, who was chosen for umpire, wou'd naturally, to shew his impartiality give every one the product of his own country. (Hume (1740) 2000, 3.2.3: 10 n. 5)

The German, the Frenchman, and the Spaniard each presumably would like to drink all three bottles of wine. So they are prone to making incompatible claims on this limited quantity of wine. Like Braithwaite in his analysis of Matthew and Luke's predicament, Hume in presenting his Wine Division Problem adopts the perspective of an arbiter who will choose a division of the goods at stake on behalf of the claimants. Hume's problem differs from the other division problems discussed above in that there are more than two claimants. In their original presentations, Nash and Raiffa focus on bargaining problems with two claimants, but bargaining problems with three or more claimants present new levels of complexity.

Another interesting way Hume's problem differs from the Chocolate Cake and Neighboring Musician problems is that in Hume's problem the goods at stake are not clearly homogeneous. A partition of the wine could assign quantities of three different vintages to each claimant. In this respect Hume's problem resembles Nash's Trading Problem where Bill and Jack have sets of different toys available for exchange. Hume in fact exploits the fact that there are different vintages of wines in his problem. The solution Hume proposes assigns each claimant one bottle, the bottle containing the wine of the vintage produced in the claimant's home country. Hume's solution defines a partition of the wine that can characterize a small-scale Humean convention for this society of three. There are of course infinitely many distinct possible partitions of the wine. Any partition that allots a positive quantity of wine to each of the three claimants characterizes a strict Nash equilibrium, since each would get less wine if he were to claim less than his share of this partition, and no wine at all if he were to claim more than his share of this partition and thereby prolong their quarrel. Given so many alternative equilibria, why should Hume's arbiter implement Hume's solution? And if Hume's solution is the "right" solution, then why do the three claimants need the arbiter's help?

---

knowledge of a proposition *A* for a group *N* implies that each member of *N* has reason to believe that *A* obtains, each member of *N* has reason to believe that each member of *N* has reason to believe that *A* obtains, and so on.

3 Defining the Problem

Throughout this Element, the set $N = \{1, \ldots, n\}$ will be used to refer to the n agents engaged in a bargaining problem or game where each $i \in N$ designates a unique agent of N, also referred to as *Agent i*. An n-agent *(Nash) bargaining problem* for the agents of N is an ordered pair $(S, \boldsymbol{u}_0)$ where S, the *feasible set*, is a subset of the space $\mathbb{R}^n$ of points that are n-tuples of real numbers and where $\boldsymbol{u}_0 = (u_{01}, \ldots, u_{0n}) \in S$, the *nonagreement point*, is one of these n-tuples.[9] A bargaining problem $(S, \boldsymbol{u}_0)$ has a two-part standard interpretation: (i) a given point or *payoff vector* $\boldsymbol{u} = (u_1, \ldots, u_n) \in S$ defines each Agent i's *payoff* u_i, understood as a utility level, if the n agents all together agree to follow some joint course of action that results in their achieving $\boldsymbol{u} \in S$; and (ii) if they fail to agree upon any such course of action, then each Agent i gets u_{0i}, Agent i's part of the nonagreement point $\boldsymbol{u}_0$. Some solution concepts for the bargaining problem presuppose the agents' payoffs are *von Neumann–Morgenstern utilities* ((1944) 2004, 617–632), each of which is uniquely defined only up to an individual choice of scale, while others presuppose that these payoffs are in some sense comparable across agents. If a point $\boldsymbol{u} \in \mathbb{R}^n$ lies outside of S – that is, $\boldsymbol{u} \notin S$ – then $\boldsymbol{u}$ is unreachable in the sense that the agents cannot achieve the joint payoffs that $\boldsymbol{u}$ would define, no matter how they might try to collaborate.

3.1 Bargaining Problems Based upon the Motivating Examples

The following are initial illustrations based upon the Section 2 examples.

Chocolate Cake Problem

Figure 1 depicts the bargaining problem $(T, (-1, -1))$ of an instance of the Chocolate Cake example. Claudia's and Laura's respective utilities for the claim profile $\boldsymbol{x} = (x_1, x_2)$ where Claudia claims $x_1 \in [0, 1]$ and Laura claims $x_2 \in [0, 1]$ are $u_1(\boldsymbol{x}) = x_1$ and $u_2(\boldsymbol{x}) = x_2$ when $x_1 + x_2 \leq 1$ – that is, when their claims x_1 and x_2 are compatible; and $u_1(\boldsymbol{x}) = u_2(\boldsymbol{x}) = -1$ when their claims are incompatible. In this figure, and in all similar figures, the light shaded region depicts the feasible set and the circled point depicts the nonagreement point. The feasible set T is the triangular region in $\mathbb{R}^2$ with vertices $(1, 0)$, $(0, 1)$, and $(-1, -1)$ and the nonagreement point is $\boldsymbol{u}_0 = (-1, -1)$. For this Chocolate Cake Problem Claudia's and Laura's utility functions are both linear and each increases monotonically in the amount of cake each receives, regardless of how

[9] The feasible set S has other names in the literature, including *negotiation set* and *bargaining region*. The nonagreement point $\boldsymbol{u}_0$ also has other names in the literature, including *status quo point* and *conflict point*.

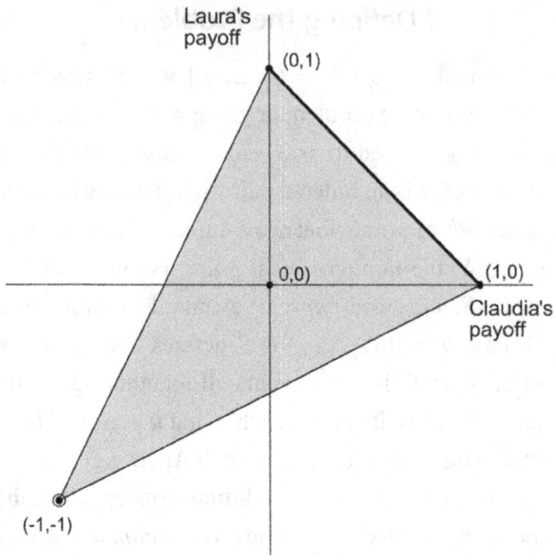

Figure 1 Simple Chocolate Cake bargaining problem

much cake the other receives. Here the nonagreement point reflects an outcome where neither receives any cake and they fight, the worst possible outcome for both of them.

Trading Problem

Figure 2 depicts the bargaining problem $(R, (12, 6))$ of Nash's Trading Problem. In this bargaining problem the nonagreement point is $\boldsymbol{u}_0 = (12, 6)$ defined by the outcome where Bill and Jack make no trade, and the payoff vectors of the feasible set R are defined by all of the possible lotteries over the 512 possible deterministic trades. Nash made no explicit reference to these lotteries when he introduced this example, but it will become evident below that he implicitly assumed that Bill and Jack can trade parts of these lotteries as well as make deterministic trades.

Braithwaite Problem

Figure 3 depicts the bargaining problem $(B, (0, \frac{1}{9}))$ of Braithwaite's neighboring musicians.[10] In the Braithwaite Problem the feasible set B is the polygonal region in $\mathbb{R}^2$ with vertices $(0, \frac{1}{9})$, $(\frac{1}{6}, 0)$, $(1, \frac{2}{9})$, and $(\frac{1}{2}, 1)$ and the nonagreement point is $\boldsymbol{u}_0 = (0, \frac{1}{9})$. Luke's and Matthew's respective utilities at the claim pair $\boldsymbol{x} = (x_1, x_2)$ given their respective claims x_1 and x_2 on the amount of available

[10] Here the payoffs are those of one of Luce and Raiffa's scalings in their discussion of the Braithwaite Problem (1957, 146–147).

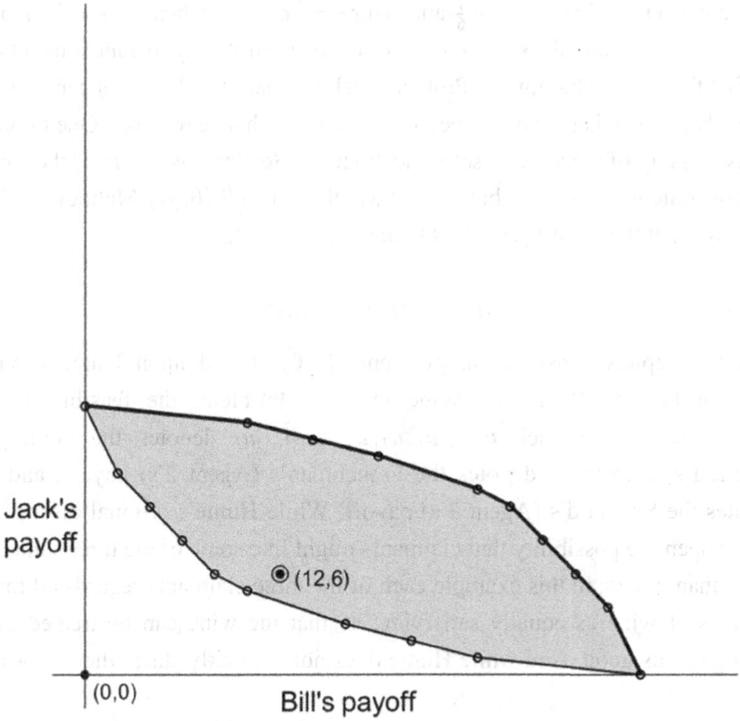

Figure 2 Nash's Trading Problem

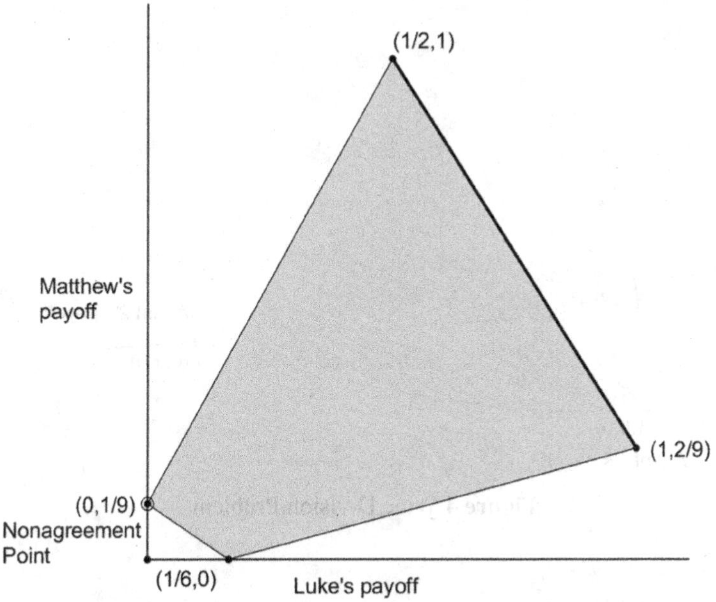

Figure 3 Braithwaite Problem

time are $u_1(\boldsymbol{x}) = \frac{5}{6}x_1 - \frac{1}{3}x_2 + \frac{1}{6}$ and $u_2(\boldsymbol{x}) = \frac{2}{9}x_1 + x_2$ when $x_1 + x_2 \leq 1$ and $u_1(\boldsymbol{x}) = u_{01} = 0$ and $u_2(\boldsymbol{x}) = u_{02} = \frac{1}{9}$ otherwise. These payoff functions reflect the fact that in the Braithwaite Problem each claimant's utility level can depend somewhat upon what the other receives as well as what he receives. One important asymmetry of the feasible set B mentioned in Section 2 is that $\boldsymbol{u}_0$ is the worst possible outcome for Luke but not for Matthew. In $\left(B, \left(0, \frac{1}{9}\right)\right)$ Matthew in fact does worse at the point $\left(\frac{1}{6}, 0\right) \in B$ than at $\boldsymbol{u}_0 = \left(0, \frac{1}{9}\right)$.

Wine Division Problem

Figure 4 depicts a bargaining problem $(W, \mathbf{0}_3)$ based upon Hume's Wine Division Problem.[11] In this Wine Division Problem, the feasible set is $W \subset \mathbb{R}^3$ where for each $\boldsymbol{u} = (u_1, u_2, u_3) \in W$, u_1 denotes the German's (Agent 1's) payoff, u_2 denotes the Frenchman's (Agent 2's) payoff, and u_3 denotes the Spaniard's (Agent 3's) payoff. While Hume's original discussion leaves open the possibility that claimants might like some of the three vintages better than others, in this example each of the three claimants regards all three vintages of wine as equally satisfying, so that the wine can be treated as a homogeneous good. And while Hume does not explicitly state whether or not

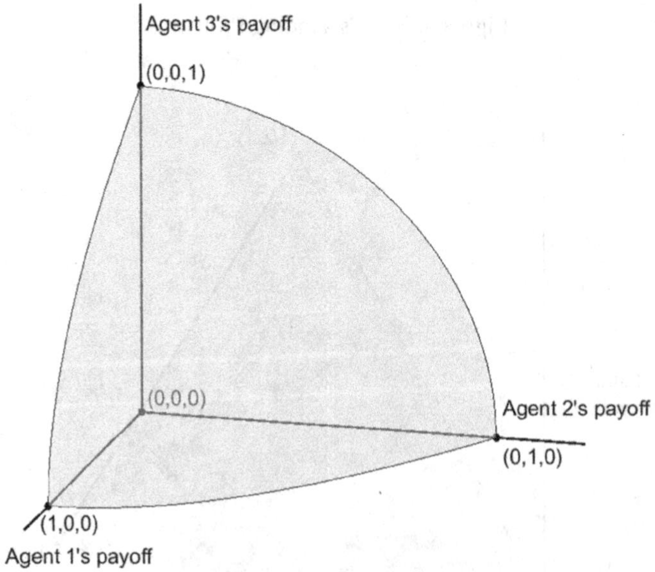

Figure 4 Wine Division Problem

[11] The payoff functions of this example are due to Young 1994, 121–122.

the claimants bear additional costs for continuing their quarrel or whether or not any of them care how much wine either of the other two receives, in this example each claimant's preferences over all possible outcomes depend only upon the amount of wine he receives. For this problem, the nonagreement point is $\boldsymbol{u}_0 = \mathbf{0}_3 = (0, 0, 0)$. If the German claims $x_1 \in [0, 1]$ of the available wine, the Frenchman claims $x_2 \in [0, 1]$ of the available wine, and the Spaniard claims $x_3 \in [0, 1]$ of the available wine, then at any profile of compatible claims $\boldsymbol{x} = (x_1, x_2, x_3)$ where $x_1 + x_2 + x_3 \leq 1$, their respective payoffs are $u_1(\boldsymbol{x}) = x_1$, $u_2(\boldsymbol{x}) = (x_2)^{\frac{1}{2}}$, and $u_3(\boldsymbol{x}) = (x_3)^{\frac{1}{2}}$, and otherwise $u_1(\boldsymbol{x}) = u_2(\boldsymbol{x}) = u_3(\boldsymbol{x}) = 0$. The German's linear payoff function reflects his increasing and constant marginal utility for more wine, while the Frenchman's and Spaniard's payoff functions reflect their increasing but diminishing marginal utilities for more wine. The curved surface that is part of the boundary of W is defined by the payoff vectors $(u_1(\boldsymbol{x}), u_2(\boldsymbol{x}), u_3(\boldsymbol{x}))$ where $x_1 + x_2 + x_3 = 1$ – that is, the available wine is partitioned and distributed among the three claimants. Here the nonagreement point corresponds to the outcome where the three claimants continue their quarrel and none gets any wine. Given these payoff functions, $\boldsymbol{u}_0 = \mathbf{0}_3$ is a worst possible outcome for each of the claimants, but any other outcome where one or two of the claimants get none of the wine is also a worst possible outcome for him or them.

3.2 Convexity and Compactness

Most discussions of the Nash bargaining problem assume that the feasible set S satisfies certain structural properties, including in particular that S is *convex* and *compact*. A convex combination of points is a weighted average of these points with nonnegative weights that sum to one. A set Λ is convex when any convex combination of two points in this set is again in Λ, or equivalently when the line segment joining these two points lies entirely in Λ. A set Λ is compact when Λ is bounded and contains its own boundary. The feasible sets of the Figure 1, Figure 2, Figure 3, and Figure 4 bargaining problems are all convex and compact. Convexity reflects two ideas. The first of these ideas is that in a bargaining problem the goods at stake are infinitely divisible – if not physically so, then by permitting divisions of parts of lotteries over these goods. The second of these ideas is that the utility function of each claimant increases in received shares of goods with constant or diminishing marginal utility. Compactness reflects the ideas that the goods of a bargaining problem can be divided exactly, with nothing left over, and that there are definite limits both to how much the agents involved can benefit by collaborating and how much harm they suffer if they fail to collaborate.

Some of the significance of the convexity and compactness properties can be illustrated with counterexamples. In his trading example, Nash specifies that the feasible set of Bill and Jack's bargaining problem is convex (1950, 161). This makes sense if one assumes Bill and Jack can trade parts of lotteries over their 512 possible deterministic trades. But if Jack and Bill are only capable of making deterministic trades, then they can achieve only one of the payoff vectors in the finite set depicted in Figure 5. The Figure 5 feasible set is a proper subset of the convex Figure 2 feasible set, and is not itself convex since in fact none of the line segments connecting any of the pairs of points in this set is itself in this set. The Figure 2 feasible set is a convex polygon that extends the Figure 5 feasible set by relaxing the assumption that Jack and Bill are limited to deterministic trades and permitting trades of parts of lotteries over the deterministic trades.

Figure 6 depicts the feasible set of a twist on the Figure 1 Chocolate Cake bargaining problem where $u_1(x_1, x_2) = x_1$ and $u_2(x_1, x_2) = x_2$ when $x_1 + x_2 < 1$ and $u_1(x_1, x_2) = u_2(x_1, x_2) = -1$ otherwise. Now Claudia and Laura each receives the share she claims, on condition there is more than enough cake to provide each with her claimed share, with no other restriction. They must leave some positive quantity of cake unclaimed in order for each of them to get any cake at all, but there is no lower bound upon this unclaimed quantity. In this bargaining problem the feasible set is not compact. The payoff

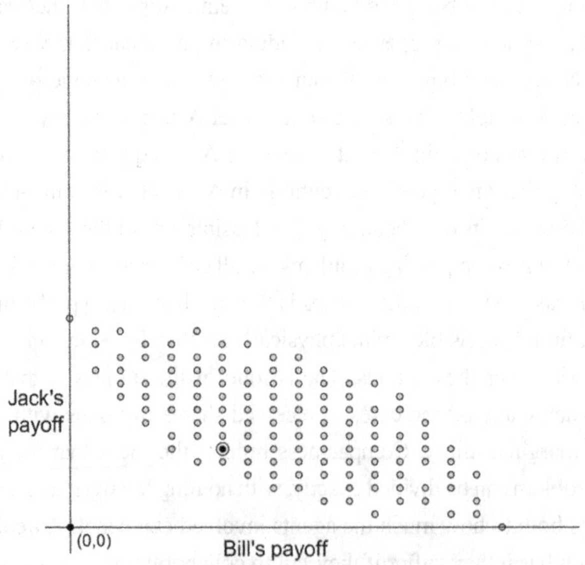

Figure 5 Feasible set for Nash Trading Problem with deterministic trades only

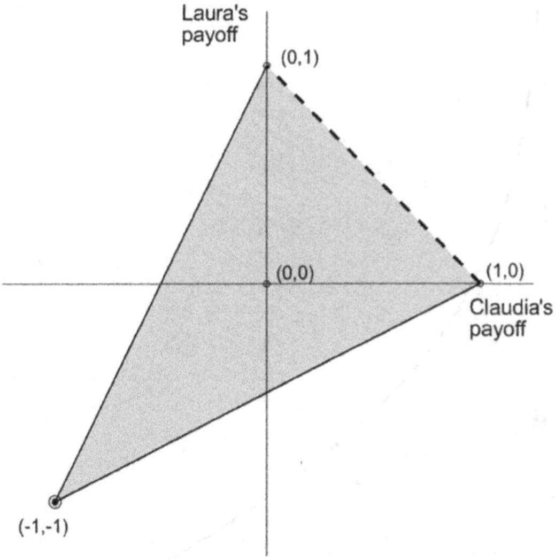

Figure 6 A feasible set that does not contain all of its own boundary

vectors of this feasible set include all of the triangular region of the Figure 1 feasible set up to but not including its northeast boundary, for these boundary points correspond to claim parts that divide the cake exactly, with nothing left over. Laura and Claudia cannot reach any of these boundary points in this example because any exact division of the cake is effectively prohibited.

Figure 7 summarizes a different Chocolate Cake bargaining problem where the nonagreement point is now $\boldsymbol{u}_0 = \boldsymbol{0}_2 = (0,0)$, and this time $u_1(x_1, x_2) = x_1$ and $u_2(x_1, x_2) = x_2 - \log(x_1)$ when claims are compatible and $u_1(x_1, x_2) = u_2(x_1, x_2) = 0$ otherwise. In this example the feasible set Q is neither compact nor convex, since Q is not bounded and many convex combinations of point pairs in Q do not lie fully in Q. One can interpret Laura's utility function in this example that leads to Q being unbounded as her drawing some malicious satisfaction from the outcomes of certain pairs of compatible claims that increase without bound as Claudia's share decreases.

3.3 Dominance-Based Feasible Set Properties

Many discussions of bargaining problems assume that in addition to convexity and compactness, the feasible set S satisfies some additional properties based upon dominance relations across payoff vectors. Moreover, bargaining theory in general makes frequent use of *Pareto optimality* concepts, which are also based upon these dominance relations. Here I will first review some of these dominance relations and Pareto optimality concepts. Then I will discuss the

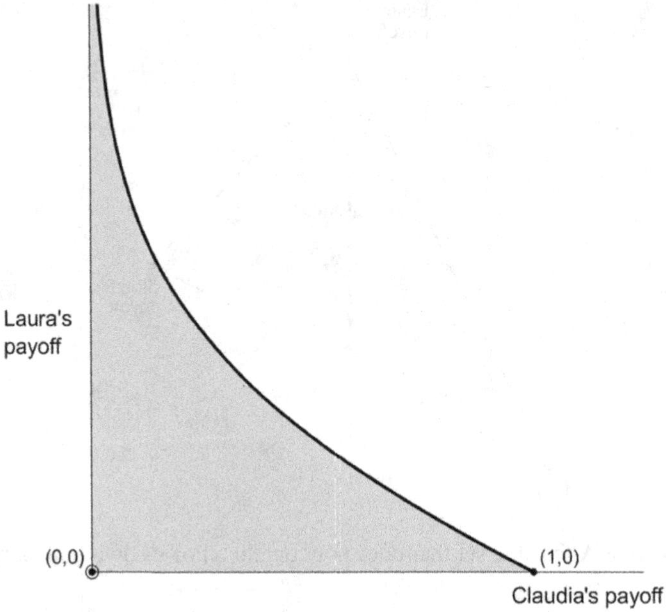

Figure 7 An unbounded and nonconvex feasible set

dominance-based structural properties often associated with bargaining problem feasible sets.

The points that are ordered n-tuples of real numbers can be partially ordered in the following way: For points $\boldsymbol{u} = (u_1, \ldots, u_n)$ and $\boldsymbol{v} = (v_1, \ldots, v_n)$, $\boldsymbol{u} \geqq \boldsymbol{v}$ when $u_i \geq v_i$ for each $i \in N$. Point $\boldsymbol{u}$ *weakly dominates* point $\boldsymbol{v}$, written $\boldsymbol{u} \geq \boldsymbol{v}$, when $\boldsymbol{u} \geqq \boldsymbol{v}$ and $\boldsymbol{u} \neq \boldsymbol{v}$ – that is, when $u_i \geq v_i$ for each $i \in N$ and $u_i > v_i$ for at least one $i \in N$. Point $\boldsymbol{u}$ *strictly dominates* $\boldsymbol{v}$, written $\boldsymbol{u} > \boldsymbol{v}$, when $u_i > v_i$ for each $i \in N$. Strict dominance is a special case of weak dominance. When one interprets points as payoff vectors, then: (i) When $\boldsymbol{u} \geq \boldsymbol{v}$ then at $\boldsymbol{u}$ each agent fares at least as well as she fares at $\boldsymbol{v}$ and some agents fare better at $\boldsymbol{u}$ than at $\boldsymbol{v}$; and (ii) If $\boldsymbol{u} > \boldsymbol{v}$ then every agent fares better at $\boldsymbol{u}$ than at $\boldsymbol{v}$. For a given set Λ, a point $\boldsymbol{u} \in \Lambda$ is *weakly Pareto optimal* in Λ if no point in Λ strictly dominates $\boldsymbol{u}$ or, equivalently, for each $\boldsymbol{v} \in \mathbb{R}^n$ if $\boldsymbol{v} > \boldsymbol{u}$ then $\boldsymbol{v} \notin \Lambda$. The *undominated boundary* P^0_Λ of Λ is the set of weakly Pareto optimal points in Λ. $\boldsymbol{u} \in \Lambda$ is *Pareto optimal* in Λ if no point in Λ weakly dominates $\boldsymbol{u}$ or, equivalently, for each $\boldsymbol{v} \in \mathbb{R}^n$ if $\boldsymbol{v} \geq \boldsymbol{u}$ then $\boldsymbol{v} \notin \Lambda$. The *Pareto frontier* P^+_Λ of Λ is the set of Pareto optimal points in Λ. P^0_Λ and P^+_Λ are nonempty sets when Λ is compact and convex, and $P^+_\Lambda \subseteq P^0_\Lambda$ since Pareto optimality implies weak Pareto optimality for any set. For both the Figure 1 and Figure 3 bargaining problems the Pareto frontier is the northeast boundary of the feasible set. P^+_T and P^+_B each consists of the payoff vectors of the form $(u_1(\boldsymbol{x}), u_2(\boldsymbol{x}))$ where $x_1 + x_2 = 1$ corresponding to exact divisions of the good at stake. In these figures, and in all similar figures for two-agent bargaining problems, the bold-faced curve of the boundary of S depicts the Pareto frontier P_S^+. For the three-agent Figure 4 bargaining problem, P_W^+ is the set of points that defines the curved surface described above consisting of the payoff vectors $(u_1(\boldsymbol{x}), u_2(\boldsymbol{x}), u_3(\boldsymbol{x}))$ where $x_1 + x_2 + x_3 = 1$.

A bargaining problem $(S, \boldsymbol{u}_0)$ is *essential* when S is compact and convex and contains at least one point $\boldsymbol{v} > \boldsymbol{u}_0$. Essentiality reflects the idea that the agents can all gain together by collaborating. David Gauthier introduces a piece of terminology helpful for expressing this idea when he argues that rational contractors are concerned with distributing the *cooperative surplus* – that is, the set of gains they can achieve beyond their initial bargaining position payoffs (Gauthier 1986, 130, 146). One can identify the *cooperative surplus vectors* of an essential $(S, \boldsymbol{u}_0)$ as the subset $S^+ \subset S$ of payoff vectors that weakly dominate $\boldsymbol{u}_0$.[12] The dark shaded region of Figure 8 depicts B^+ for the Braithwaite Problem. The dark shaded region of Figure 9 depicts R^+ for the Nash Trading Problem.

An n-agent bargaining problem $(S, \boldsymbol{u}_0)$ is *nontrivial* when S contains a set of n distinct points $\boldsymbol{v}_1, \ldots, \boldsymbol{v}_n$ such that: (i) $\boldsymbol{v}_i = (v_{i1}, \ldots, v_{in}) \geqq \boldsymbol{u}_0$, (ii) $v_{ii} > u_{0i}$, and (iii) $v_{ii} > v_{ji}$ for $i \neq j$. Conditions (i) and (ii) imply that each $\boldsymbol{v}_i$ weakly dominates the nonagreement point $\boldsymbol{u}_0$ and that Agent i fares strictly better at $\boldsymbol{v}_i$ than at $\boldsymbol{u}_0$. Condition (iii) implies that if the agents follow any $\boldsymbol{v}_j \neq \boldsymbol{v}_i$, then Agent i fares less well than she fares at $\boldsymbol{v}_i$. Nontriviality is a special case of essentiality,[13] and reflects the idea that while the agents can all gain by collaborating, they have some conflicting preferences over how they might collaborate. The Figure 1, Figure 2, Figure 3 and Figure 4 bargaining problems are all nontrivial. All of the bargaining problems discussed in this Element will be assumed to be essential, and unless expressly stated otherwise will also be assumed to be nontrivial.

---

12 In his essay "Bargaining and Justice," Gauthier includes essentiality in his definition of a bargaining problem and defines his preferred solution concept in terms of S^+ ((1985), 30, 36). In his definition of the cooperative surplus in *Morals by Agreement*, Gauthier only explicitly requires that for each $\boldsymbol{u} \in S^+, \boldsymbol{u} \geqq \boldsymbol{u}_0$ (1986, 141). So strictly speaking, according to Gauthier's definition $\boldsymbol{u}_0$ would be one of the cooperative surplus vectors and a nonessential bargaining problem would have a nonempty set of cooperative surplus payoff vectors, namely $\{\boldsymbol{u}_0\}$ only. I think the definition here that excludes the nonagreement point does no serious violence to Gauthier's contributions and may even reflect what Gauthier has in mind given his discussion in "Bargaining and Justice."

13 For a convex combination $\boldsymbol{v}^* = x_1\boldsymbol{v}_1 + \cdots + x_n\boldsymbol{v}_n$ where each $x_i > 0$, $\boldsymbol{v}^* = (v_1^*, \ldots, v_1^*) \in S$ because S is convex. Moreover, since $v_{ji} \geq u_{0i}$ for each Agent j by condition (i), conditions (i) and (ii) together imply that $v_i^* = x_1 v_{i1} + \cdots + x_n v_{in} > u_{i0}$. Hence $\boldsymbol{v}^* > \boldsymbol{u}_0$ and so $(S, \boldsymbol{u}_0)$ is essential.

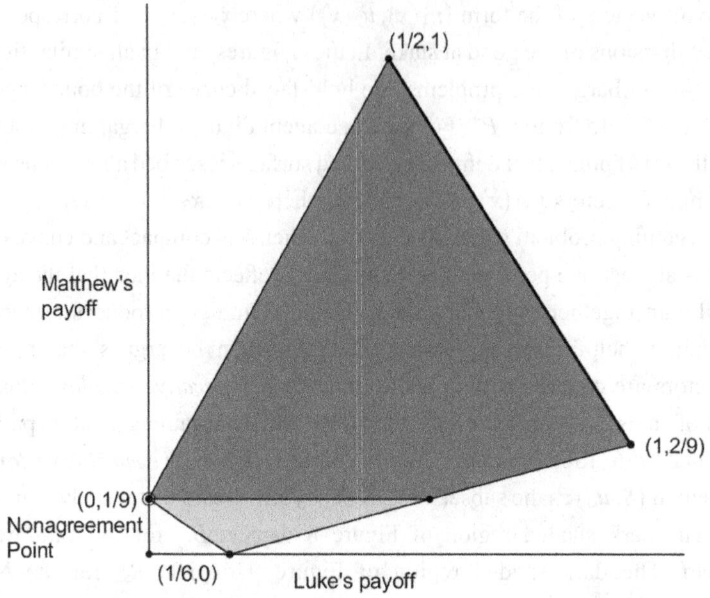

Figure 8 Cooperative surplus vectors of the Braithwaite Problem

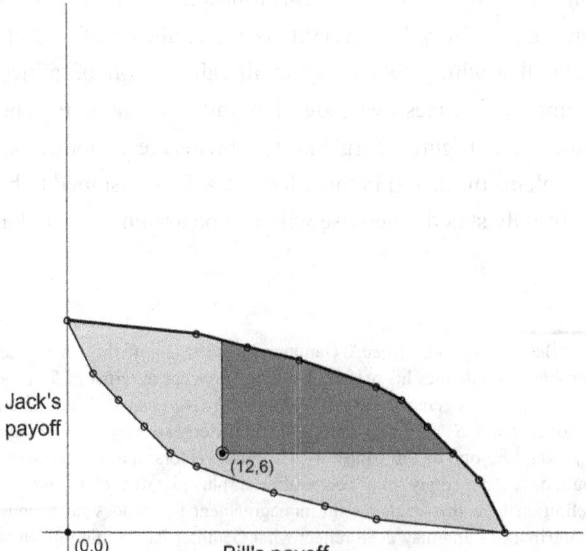

Figure 9 Cooperative surplus vectors of the Nash Trading Problem

Two additional *comprehensiveness* properties play important roles in the analysis of many bargaining problems. A bargaining problem $(S, \boldsymbol{u}_0)$ is $\boldsymbol{u}_0$*-comprehensive* when if $\boldsymbol{u} \in S^+$ and $\boldsymbol{u} \geq \boldsymbol{v} \geq \boldsymbol{u}_0$, then $\boldsymbol{v} \in S^+$. $(S, \boldsymbol{u}_0)$ is *strictly* $\boldsymbol{u}_0$*-comprehensive* when if $\boldsymbol{u}, \boldsymbol{v} \in S^+$ and $\boldsymbol{u} \geq \boldsymbol{v}$, then there exists $\boldsymbol{w} \in S^+$ such that $\boldsymbol{w} > \boldsymbol{v}$. The Figure 2 Nash Trading Problem is strictly $(12, 6)$-comprehensive. Likewise the Figure 4 Wine Division Problem is strictly $\mathbf{0}_3$-comprehensive. However, the Figure 1 Chocolate Cake Problem is not $(-1, -1)$-comprehensive and the Figure 3 Braithwaite Problem is not $(0, \frac{1}{9})$-comprehensive. A number of important results in the analysis of bargaining problems obtain only for $\boldsymbol{u}_0$-comprehensive or strictly $\boldsymbol{u}_0$-comprehensive bargaining problems. Indeed, some authors limit their analysis to bargaining problems satisfying one or both of these comprehensiveness properties. I believe that it is natural to assume comprehensiveness in some applications but not in general. When a bargaining problem $(S, \boldsymbol{u}_0)$ is $\boldsymbol{u}_0$-comprehensive, it becomes possible for any individual agent to enjoy their maximum possible gain within the cooperative surplus set S^+ while the others enjoy no gain at all. I think $\boldsymbol{u}_0$-comprehensiveness is a natural assumption in situations where: (i) the utility for each agent of failing to reach agreement is the same as if they had simply left the interaction with what they started, and (ii) the agents' preferences are all nontuistic – that is, each agent takes no interest in the preferences of the other agents. The Figure 2 and Figure 4 bargaining problems based upon Nash's and Hume's stories reflect such situations. In the Figure 2 problem, the nonagreement point payoffs are equivalent to the outcome where Bill and Jack simply make no trade, and the utilities for each depend only upon the toys he has in the end, so their preferences are nontuistic. In the Wine Division Problem the nonagreement point utilities are simply the same as they are in the event that none of the three claims any of the wine. And each claimant's utility increases according to any positive amount of wine he receives regardless of what the others receive, reflecting their nontuistic preferences. But in the Figure 1 bargaining problem, the nonagreement point leaves each claimant worse off than they would be by claiming nothing. This reflects the Section 2 Chocolate Cake story, where Claudia and Laura both regard fighting as even worse than receiving no cake but at least avoiding a fight. In the Figure 3 problem, the nonagreement point payoffs are not the same as the payoffs of the outcome where each side claims none of the playing time. Indeed, Matthew's and Luke's preferences over these two outcomes diverge. And in this problem, the utilities for Matthew and Luke each depend to some extent upon how much playing time the other receives, as well as how much playing time they themselves receive. Geometrically, this is reflected partly by the fact that the points $(1, \frac{2}{9})$ and $(\frac{1}{2}, 1)$, respectively most favorable to Luke and to Matthew, both strictly dominate

$\boldsymbol{u}_0 = \left(0, \frac{1}{9}\right)$. If Luke's and Matthew's preferences were nontuistic, then the feasible set points respectively most favorable to each would each only weakly dominate $\boldsymbol{u}_0$. These examples show that $\boldsymbol{u}_0$-comprehensiveness is not always a property that follows so automatically from the nature of the problem that generates the feasible set.

In general one can extend a noncomprehensive bargaining problem into a $\boldsymbol{u}_0$-comprehensive bargaining problem. However, how one interprets such an extension can underscore why some situations are best modeled as noncomprehensive bargaining problems. To illustrate, one can extend the Braithwaite Problem $\left(B, \left(0, \frac{1}{9}\right)\right)$ by defining a new feasible set B' where $\boldsymbol{u} \in B'$ exactly when $\boldsymbol{u} \in B$ or $x_1\left(0, \frac{1}{9}\right) + x_2(0, 1) + x_3\left(\frac{1}{2}, 1\right) + x_4\left(1, \frac{1}{9}\right) \geqq \boldsymbol{u} \geqq \left(0, \frac{1}{9}\right)$ for some $x_1, x_2, x_3, x_4 \in [0, 1]$ and $x_1 + x_2 + x_3 + x_4 = 1$, as depicted in Figure 10. The new feasible set B' contains all of B, here light shaded, of the original Braithwaite Problem together with two new light shaded triangular regions. The Figure 10 bargaining problem $\left(B', \left(0, \frac{1}{9}\right)\right)$ is $\left(0, \frac{1}{9}\right)$-comprehensive, though not strictly $\left(0, \frac{1}{9}\right)$-comprehensive. For this extended bargaining problem the Pareto frontier P_S^+ is a proper subset of the undominated boundary P_S^0, which includes the line segment joining $(0, 1)$ and $\left(\frac{1}{2}, 1\right)$ and the line segment joining $\left(1, \frac{1}{9}\right)$ and $\left(1, \frac{2}{9}\right)$ as well as P_S^+. The standard economic interpretation of this

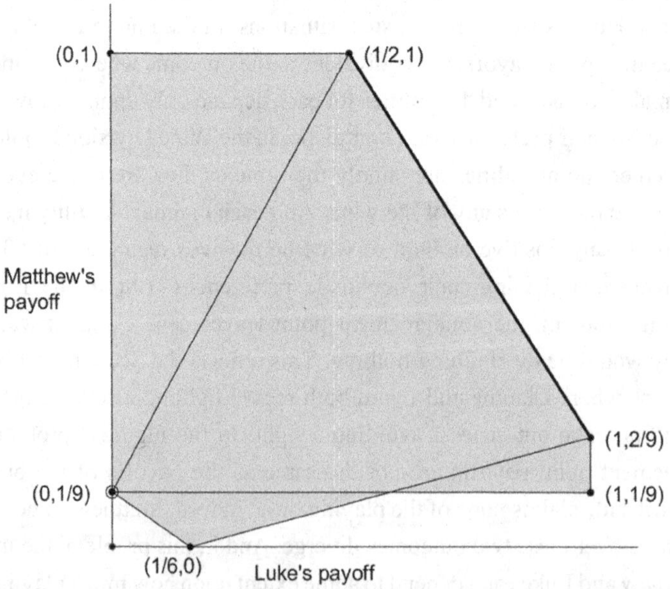

Figure 10 $\left(0, \frac{1}{9}\right)$-comprehensive extension of the Braithwaite Problem

extension is to suppose that each claimant's utility is *freely disposable* – that is, each is able to give up unilaterally some or even all of the utility he achieves by following his end of a pair of compatible claims. Many authors assume free disposability in the analyses of many concrete economic problems. If one assumes free disposability, then individual agents have the option of unilaterally "throwing away utility." But in the Braithwaite Problem, how could Luke or Matthew lower his own utility without also lowering or raising his neighbor's utility? One could add a postscript to Braithwaite's original story, according to which for any pair of compatible claims $\boldsymbol{x} = (x_1, x_2)$ that yields Luke and Matthew the payoff vector $(u_1(\boldsymbol{x}), u_2(\boldsymbol{x})) \in B$, either can unilaterally lower his utility by listening for some time $x_i' \in [0, x_i)$ by himself on earphones to some cacophony he finds as bad as the cacophony that determines his nonagreement point payoff. But one would not expect agents like Luke and Matthew, who in Braithwaite's original presentation both strictly prefer listening to their neighbor play undisturbed over enduring the cacophony that ensues when they issue incompatible claims, to each adopt some means for reducing the utility level he would achieve by simply following $\boldsymbol{x}$. More generally, one might wonder if the free disposability assumption meshes well with the even more common assumption in analyses of economic problems that individual agents are rational in the orthodox Bayesian sense of consistently choosing so as to maximize expected utility. Yet even if one takes no issue with free disposability in principle, this does not imply that one should introduce ad hoc changes into the background conditions of a situation having a bargaining problem structure just to make free disposability possible. As this example illustrates, the extension of a noncomprehensive bargaining problem into a $\boldsymbol{u}_0$-comprehensive bargaining problem may have weak underlying motivation.

In summary, comprehensiveness properties are very important in bargaining theory, and in many important applications it is natural to construct corresponding bargaining problems that satisfy these properties. But in many other important applications it is more natural to construct corresponding bargaining problems that do not satisfy these properties. And while one can formally extend any bargaining problem into a new $\boldsymbol{u}_0$-comprehensive bargaining problem, I believe the extended bargaining problem might model a situation rather different from the situation modeled by the original bargaining problem. For these reasons I do not limit the discussion of bargaining problems in this Element to those that satisfy $\boldsymbol{u}_0$-comprehensiveness or strict $\boldsymbol{u}_0$-comprehensiveness, and when these properties or their absences are important this will be addressed explicitly.

3.4 Underlying Games of Bargaining Problems

Nash formulated the bargaining problem by abstracting away from some of the specifics of the underlying interactions of the agents. The bargaining problem is defined fully in terms of payoff vectors, with details of the possible joint actions that can produce any of the available payoff vectors pushed into the background. As the Section 2 examples illustrate, the bargaining problem is a remarkably flexible framework for analyzing problems of interaction where the interests of the agents both coincide and conflict to some extent. Nash's Trading Problem illustrates one important class of such problems. One can generate the feasible set and nonagreement point of Bill and Jack's problem by constructing a function from the subsets of the goods up for trade to payoff vectors. Formally this procedure is akin to defining the utility vectors traders can achieve in an ideally competitive market. The Chocolate Cake, Braithwaite, and Wine Division Problems illustrate another important class of bargaining problems. For these resource division problems, one can build the feasible set and nonagreement point from a noncooperative *basis game*. One way to characterize such a basis game is illustrated by the discussion of the Chocolate Cake problem in Section 2, where the various claims each Agent i can issue in one of these resource division problems are Agent i's pure strategies. Viewed this way, Agent i's payoff at a given strategy profile $\boldsymbol{x}$ that is the set of agents' claims is defined by $u_i(\boldsymbol{x})$.

In some cases, an alternate approach starts with a basis game where each Agent i has finitely many pure strategies. The Figure 11 matrix summarizes such a basis game for the Figure 3 Braithwaite Problem. In this game Luke and Matthew can each follow one of two pure strategies: be *modest* (M) and claim

		Agent 2 (Matthew)	
		M	G
Agent 1 (Luke)	M	$(\frac{1}{6}, 0)$	$(\frac{1}{2}, 1)$
	G	$(1, \frac{2}{9})$	$(0, \frac{1}{9})$

M = modest (claim none), G = greedy (claim all)

Figure 11 Braithwaite basis game

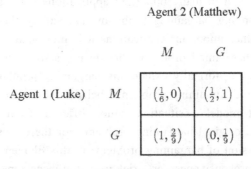

none of the good at stake, or be *greedy* (G) and claim all of this good.[14] The pure strategy profiles (G, M) and (M, G) with respective payoff vectors $\left(1, \frac{2}{9}\right)$ and $\left(\frac{1}{2}, 1\right)$ are strict Nash equilibria. If Luke and Matthew together observe the outcome $\omega \in \Omega = \{\omega_1, \omega_2\}$ of an experiment where ω_1 occurs with probability λ and ω_2 occurs with probability $1 - \lambda$, then the strategy system defined by the function

$$\boldsymbol{h}_\lambda(\omega) = \begin{cases} (G, M) \text{ if } \omega = \omega_1 \\ (M, G) \text{ if } \omega = \omega_2 \end{cases}$$

is a *strict correlated equilibrium* where Luke and Matthew alternate between the strict Nash equilibria (G, M) and (M, G) with respective probabilities λ and $1 - \lambda$. The Braithwaite basis game is a *conflictual coordination game*, since there are several strict equilibria over which the agents might coordinate and their preferences over these equilibria conflict.[15] For the corresponding bargaining problem, the nonagreement point $\boldsymbol{u}_0 = \left(0, \frac{1}{9}\right)$ is the payoff vector of the outcome (G, G) where Luke and Matthew are both greedy. One builds the feasible set S by taking all of the convex combinations over the set $\left\{\left(0, \frac{1}{9}\right), \left(\frac{1}{6}, 0\right), \left(1, \frac{2}{9}\right), \left(\frac{1}{2}, 1\right)\right\}$ of payoff vectors defined by all four pure strategy profiles of the Figure 11 game. Any convex combination of a subset of the pure strategy profiles is equivalent to a joint randomization over these profiles, and these convex combinations also define the outcomes where the claimants can claim fractional shares of the good. The correlated equilibria of the form $\boldsymbol{h}_\lambda$ for $\lambda \in [0, 1]$ characterize the Pareto frontier of this feasible set. For each $\lambda \in [0, 1]$ the correlated equilibrium $\boldsymbol{h}_\lambda$ constructed from the Figure 11 basis game is equivalent to the Nash equilibrium $(\lambda, 1 - \lambda)$ of a corresponding basis game where Luke's and Matthew's respective pure strategies are to claim shares $x_1, x_2 \in [0, 1]$, and Luke claims $x_1 = \lambda$ and Matthew claims $x_2 = 1 - \lambda$. One can often adopt either a correlated or a Nash equilibrium perspective for identifying the Pareto fronter points of a bargaining problem constructed from a basis game.

Nash recognized that one can in general construct a noncooperative game from a bargaining problem in a manner similar to the basis game approach illustrated in the previous paragraph. Nash's idea was to allow the agents

[14] In line with how the payoff vectors of a feasible set define utility levels for the n claimants of a bargaining problem, in an n-agent game, at each strategy profile outcome determined by the agents' individual strategy choices, the nth agent's payoff is the nth component of the corresponding payoff vector. For example, at the (G, M) outcome of the Figure 11 game, Luke's payoff is 1 and Matthew's payoff is $\frac{2}{9}$.

[15] Diana Richards first suggested this terminology. See Vanderschraaf and Richards 1997. Many conflictual coordination games are known as *Battle of the Sexes* games (Luce and Raiffa 1957, 90–93) or as *Hawk-Dove* games (Maynard Smith 1982, 11–12).

engaged in a bargaining problem $(S, \boldsymbol{u}_0)$ to claim specific payoffs directly, and then to have each agent receive the payoff they claim if their joint claims do not weakly dominate any point in P_S^0 and their nonagreement point payoff otherwise (1953, 36–38). In this version of a Nash demand game, each Agent i's pure strategies become the alternative payoffs Agent i can attain in S. And in such a game, the claims that define a payoff vector $\boldsymbol{u} = (u_1, \ldots, u_n)$ are compatible when $\boldsymbol{v} \geqq \boldsymbol{u}$ for some $\boldsymbol{v} \in S$, and a point $\boldsymbol{u}^* = (u_1^*, \ldots, u_n^*) \geqq \boldsymbol{u}_0$ in the set of compatible claims vectors that is not weakly dominated by some other point in this set becomes a Nash equilibrium. For if any Agent i deviates unilaterally from $\boldsymbol{u}^*$ by claiming some $u_i' \in (u_{0i}, u_i^*)$, at the resulting vector $\boldsymbol{u}' = (u_i', \boldsymbol{u}_{-i}^*) = (u_1^*, \ldots, u_{i-1}^*, u_i', u_{i+1}^*, \ldots, u_n^*)$ then Agent i's payoff is $u_i' < u_i^*$; and if Agent i deviates unilaterally from $\boldsymbol{u}^*$ by claiming some $u_i'' > u_i^*$, then the resulting vector $\boldsymbol{u}'' = (u_i'', \boldsymbol{u}_{-i}^*)$ weakly dominates $\boldsymbol{u}^*$, and the claims are no longer compatible and Agent i's payoff is $u_{0i} < u_i^*$. As Nash clearly recognized, if the original feasible set S is not already $\boldsymbol{u}_0$-comprehensive this construction automatically expands S into a $\boldsymbol{u}_0$-comprehensive set. Nash's construction bypasses the device of a basis game for use in generating a feasible set and one can apply this construction to any bargaining problem $(S, \boldsymbol{u}_0)$. Nevertheless, the basis game construction can prove valuable for gaining insights into the natures of many specific bargaining problems, especially problems where one might not want to automatically assume comprehensiveness. In the sequel both basis games and the feasible set vectors themselves will be used to develop Nash demand games for corresponding bargaining problems.

4 Solution Concepts

Exactly how is a bargaining problem "solved"? Bargaining theorists have proposed and defended quite a variety of solution concepts for the Nash bargaining problem. In this section I will review what I consider the most important solution concepts of the bargaining problem both for bargaining theorists and for social contract theorists. Some of these solution concepts presuppose that the agents' payoffs are von Neumann–Morgenstern utilities, while others presuppose these payoffs are comparable according to some single scale. I will make it clear which of the solution concepts discussed below incorporate such interpersonally comparable utilities.

I begin by examining a solution concept that receives relatively little attention in the bargaining theory literature, though it might seem of obvious importance in social contract theory. A *Utilitarian solution* of a bargaining problem $(S, \boldsymbol{u}_0)$ is a vector in S that maximizes the sum $u_1 + \cdots + u_n$ for $(u_1, \ldots, u_n) \in S$. In Bill and Jack's Trading Problem of Figure 2, the Utilitarian solutions are the payoff

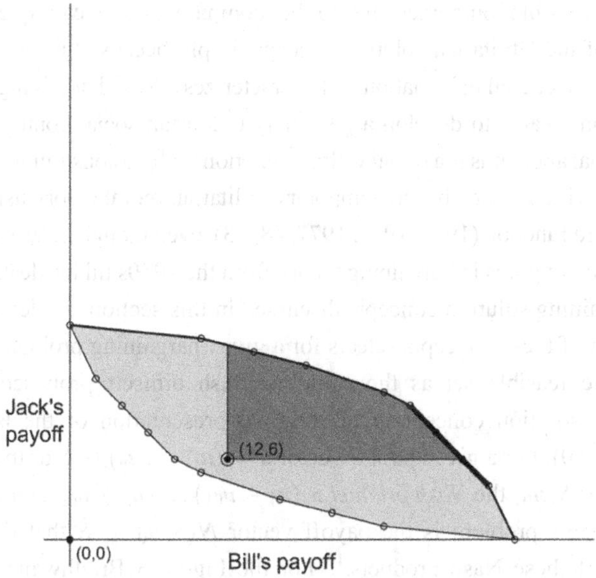

Figure 12 Utilitarian solutions of $(R, \boldsymbol{u}_0)$

vectors of the form $(u_1, u_2) \in R$ where $u_1 + u_2 = 36$. These Utilitarian solutions are depicted as the boldfaced points of the Pareto frontier of the feasible set R in Figure 12.

As its name obviously indicates, this solution concept is motivated by the utilitarian principle of maximizing the sum of expected utilities over a community of individuals. To suppose a Utilitarian solution reflects this principle in any meaningful way is to suppose interpersonally comparable utilities calibrated according to some single scale, so that the sum of the components of a payoff vector indeed measures the total utility the agents achieve at an outcome of this vector. In the Trading Problem the Utilitarian solutions are meaningful if one assumes Nash's original payoff assignments reflect Bill's and Jack's interpersonally comparable utilities according to a common scale. This problem shows that the Utilitarian solution concept is neither necessarily determinate nor mutually advantageous for the agents with respect to the nonagreement point. The feasible set R in fact has infinitely many distinct Utilitarian solutions. Moreover, some of these Utilitarian solutions fall outside the cooperative surplus set R^+, and at these Utilitarian solutions Bill is better off than at any point of R^+ while Jack is worse off than he would be at $\boldsymbol{u}_0$. The Utilitarian solution concept ignores the nonagreement point, one of the cornerstones of the Nash bargaining problem. Nash's bargaining problem is both theoretically interesting and distinctive because it makes explicit a baseline against which alternative feasible outcomes are to be compared. One can question the adequacy of the Utilitarian solution concept simply because this concept does not incorporate crucial information that characterizes a Nash bargaining problem. Indeed, if one wants to develop a genuinely utilitarian social contract, a more natural formal apparatus is a social welfare function with no constraining baseline. John Harsanyi developed his contemporary utilitarian moral theory using such a social welfare function (1953, 1955, 1977, 48–83), even though he was one of the most important figures in bargaining theory from the 1950s till his death in 2000.

The remaining solution concepts discussed in this section are determinate – that is, each of these concepts selects for a given bargaining problem a unique point of the feasible set as the solution. Nash himself proposed the first determinate solution concept in his original presentation of the bargaining problem (1950). For a given payoff vector $\boldsymbol{u} = (u_1, \ldots, u_n) \in S$ of the bargaining problem $(S, \boldsymbol{u}_0)$ the *Nash product* is $(u_1 - u_{01}) \ldots (u_n - u_{0n})$, and the *Nash solution* of this problem is the payoff vector $\boldsymbol{N}(S, \boldsymbol{u}_0) \in S$ that defines the maximum of these Nash products.[16] For the Figure 3 Braithwaite Problem, $\boldsymbol{N}(B, (0, \frac{1}{9})) = (\frac{15}{28}, \frac{17}{18}) \approx (0.536, 0.944)$ and the corresponding claim vector is $\boldsymbol{x}_N = (\frac{1}{14}, \frac{13}{14})$, so that in the Figure 11 Braithwaite basis game the Nash solution requires Luke and Matthew to follow the (G, M) outcome $\frac{1}{14}$ of the time and the (M, G) outcome $\frac{13}{14}$ of the time.[17] For the Figure 2 Nash Trading Problem, $\boldsymbol{N}(R, (12, 6)) = (24, 11)$.

Raiffa (1951, 1953) introduced two other solution concepts that rival Nash's solution in importance.[18] One of these solutions incorporates interpersonal utility comparisons. According to this *Egalitarian solution*,[19] one first applies some common scaling to the agents' utility functions and then chooses the point $\boldsymbol{E}(S, \boldsymbol{u}_0)$ yielding the agents their greatest equal utility gains within S from their

16 The Nash product defines a real-valued function that is continuous over compact set S and hence reaches a maximum value for some payoff vector in S. One can show that $\boldsymbol{N}(S, \boldsymbol{u}_0)$ is unique in several different ways. One way to establish uniqueness is to note that maximizing the Nash product function is equivalent to maximizing the function defined by taking the natural logarithm of the Nash product, and this natural logarithm function is a strictly concave function over S and hence has a unique maximizing argument, which is $\boldsymbol{N}(S, \boldsymbol{u}_0)$.

17 For this problem, the Nash product is maximized at a point along the Pareto frontier P_S^+, which is the line segment defined by $y = x \cdot (\frac{5}{3} - \frac{14x}{9})$ where $x \in [\frac{1}{2}, 1]$. For a point $(u_1(x_1), u_2(x_2)) = (u_1(x_1), u_2(1 - x_1)) = (x_1, \frac{5}{3} - \frac{14}{9}x_1) \in P_S^+$, the Nash product is $x_1 \cdot (\frac{16-14x_1}{9} - \frac{1}{9}) = x_1 \cdot (\frac{5}{3} - \frac{14x_1}{9})$, and this product is maximized at $x_1 = \frac{15}{28}$ so that $\boldsymbol{N}(S, (0, \frac{1}{9})) = (\frac{15}{28}, \frac{17}{18})$. Solving $z \cdot (\frac{1}{2}, 1) + (1 - z) \cdot (1, \frac{2}{9}) = (\frac{15}{28}, \frac{17}{18})$ yields $z = \frac{1}{14}$, so at the Nash solution Luke claims $\frac{1}{14}$ and Matthew claims $\frac{13}{14}$.

18 Raiffa introduced these two solution concepts for the two-agent case, but the characterizations given in this section apply to the n-agent case.

19 This solution is also known as the *Proportional solution*, in part because this is the name Ehud Kalai used when he published an early axiomatization of this concept (1977).

nonagreement point payoffs according to this scaling. So for a given bargaining problem $(S, \boldsymbol{u}_0)$ with such a scaling of payoff vectors, the Egalitarian solution $\boldsymbol{E}(S, \boldsymbol{u}_0)$ is the payoff vector in S of the form $\boldsymbol{u}_0 + (r, \ldots, r) = \boldsymbol{u}_0 + r\mathbf{1}_n$, where $\mathbf{1}_n = (1, \ldots, 1)$, that maximizes $r \geq 0$. Braithwaite ((1955) 1994) argues that the Egalitarian solution given a payoff scaling he defines in terms of Luke's and Matthew's maximin and minimax strategy payoffs is the fair solution of his neighboring musicians problem.[20] Luce and Raiffa consider a different and conceptually simpler scaling where both Luke's and Matthew's most desired and least desired outcomes have respective payoffs 1 and 0 (1957, 146–147). The Egalitarian solution of the Braithwaite Problem given Luce and Raiffa's scaling is $\boldsymbol{E}\left(B, \left(0, \frac{1}{9}\right)\right) = \left(\frac{15}{23}, \frac{158}{207}\right) \approx (0.652, 0.763)$, and the corresponding claim vector is $\boldsymbol{x}_E = \left(\frac{7}{23}, \frac{16}{23}\right)$. Given Nash's original scaling, for the Nash Trading Problem, $\boldsymbol{E}\left(R, (12, 6)\right) = \left(\frac{75}{4}, \frac{51}{4}\right) = (18.75, 12.75)$.

Another of Raiffa's proposed solutions, now known as the *Kalai–Smorodinsky solution*, is defined in terms of the nonagreement point $\boldsymbol{u}_0$ together with the *ideal point* $\overline{\boldsymbol{u}} = (\overline{u}_1, \ldots, \overline{u}_n)$ where $\overline{u}_i$ is Agent i's highest possible payoff in the cooperative surplus set S^+. Typically $\overline{\boldsymbol{u}}$ will lie outside of S.[21] The Kalai–Smorodinsky solution $\boldsymbol{K}(S, \boldsymbol{u}_0)$ is the payoff vector in S of the form $r\overline{\boldsymbol{u}} + (1 - r)\boldsymbol{u}_0$ that maximizes $r \geq 0$. At this payoff vector $\boldsymbol{K}(S, \boldsymbol{u}_0) = \left(u_1^*, \ldots, u_n^*\right)$ defined by this maximal $\overline{r}$, for each Agent i the ratio of their *settlement gain* $u_i^* - u_{0i}$ to their *ideal gain* $\overline{u}_i - u_{0i}$ is $\frac{u_i^* - u_{0i}}{\overline{u}_i - u_{0i}} = \overline{r}$.[22] If one thinks of one's *proportionate gain* at a payoff vector $\boldsymbol{v} = (v_1, \ldots, v_n) \in S$ where $\boldsymbol{v} \geq \boldsymbol{u}_0$ as the ratio $\frac{v_i - u_{0i}}{\overline{u}_i - u_{0i}}$, then the Kalai–Smorodinsky solution is the payoff vector of S that maximizes equal proportionate gains across the agents. The Kalai–Smorodinsky solution gets its name because in a now classic essay Ehud Kalai and Meir Smorodinsky (1975) proposed a set of axioms that characterize Raiffa's solution, although like Raiffa, Kalai and Smorodinsky did not claim this solution is necessarily the "correct" one. For the Braithwaite Problem, the Kalai–Smorodinsky solution is $\boldsymbol{K}\left(B, \left(0, \frac{1}{9}\right)\right) = \left(\frac{15}{22}, \frac{71}{99}\right) \approx (0.682, 0.717)$, and the corresponding claim vector is $\boldsymbol{x}_K = \left(\frac{4}{11}, \frac{7}{11}\right)$. For the Nash Trading Problem, $\boldsymbol{K}\left(R, (12, 6)\right) \approx (23.17, 11.28)$. Figure 13 depicts the Kalai–Smorodinsky solution, the Nash solution, and the Egalitarian solution for the Luce–Raiffa payoff scaling of the Braithwaite

20 Luce and Raiffa's discussion in *Games and Decisions* (1957, 148–150) remains the finest overview of Braithwaite's Egalitarian solution to his Braithwaite Problem.

21 Exceptions include certain essential bargaining problems that are trivial. For example, if S is the line segment consisting of the possible convex combinations of the points $(0, 0)$ and $(1, 1)$ and $u_0 = (0, 0)$, then $\overline{\boldsymbol{u}} = (1, 1) \in S$.

22 To see why, note that for each Agent i, $u_i^* = \overline{r} \cdot \overline{u}_i + (1 - \overline{r}) \cdot u_{0i} = \overline{r}(\overline{u}_i - u_{0i}) + u_{0i}$, so

$$\frac{u_i^* - u_{0i}}{\overline{u}_i - u_{0i}} = \frac{\left(\overline{r}(\overline{u}_i - u_{0i}) + u_{0i}\right) - u_{0i}}{\overline{u}_i - u_{0i}} = \overline{r}.$$

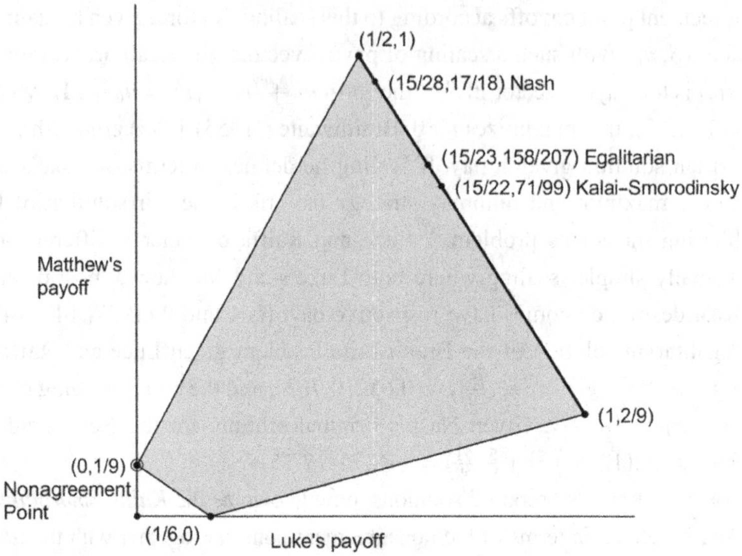

Figure 13 Three axiomatic solutions of the Braithwaite Problem

Problem. Figure 14 depicts the Kalai–Smorodinsky solution, the Nash solution, and the Egalitarian solution for the Luce–Raiffa payoff scaling of the Nash Trading Problem.

One can locate each of the Egalitarian, the Kalai–Smorodinsky, and the Nash solutions via an elementary geometric construction. These constructions are instructive in that they lend insight into why one might find each of these solutions intuitively appealing at some level. To locate the Egalitarian solution of the n-agent bargaining problem $(S, \boldsymbol{u}_0)$, one constructs the line that passes through the points $\boldsymbol{u}_0$ and $\boldsymbol{u}_0 + \mathbf{1}_n$.[23] If this line intersects S^+ then it intersects the undominated boundary P_S^0 at exactly one point, and $\boldsymbol{E}(S, \boldsymbol{u}_0)$ is this point of intersection. Otherwise $\boldsymbol{E}(S, \boldsymbol{u}_0) = \boldsymbol{u}_0$. Figure 15 illustrates this construction for the Braithwaite Problem, where the line in $\mathbb{R}^2$ passes through the points $\left(0, \frac{1}{9}\right)$ and $\left(1, \frac{10}{9}\right)$. At the point of intersection $\left(\frac{15}{23}, \frac{158}{207}\right)$ with P_B^0, which is also the northeast vertex of the largest square within S^+ with southwest vertex $\boldsymbol{u}_0$ and edges parallel to the coordinate axes, the payoffs for both agents are $\frac{15}{23}$ units greater than their respective nonagreement point payoffs. $\frac{15}{23}$ is also the length of each edge of this largest square. So by reaching this $\boldsymbol{E}\left(B, \left(0, \frac{1}{9}\right)\right)$ point these agents have achieved the best possible equal utility gains with

[23] This line is the set of points of the form $\boldsymbol{u}_0 + t\mathbf{1}_n$, $t \in \mathbb{R}$.

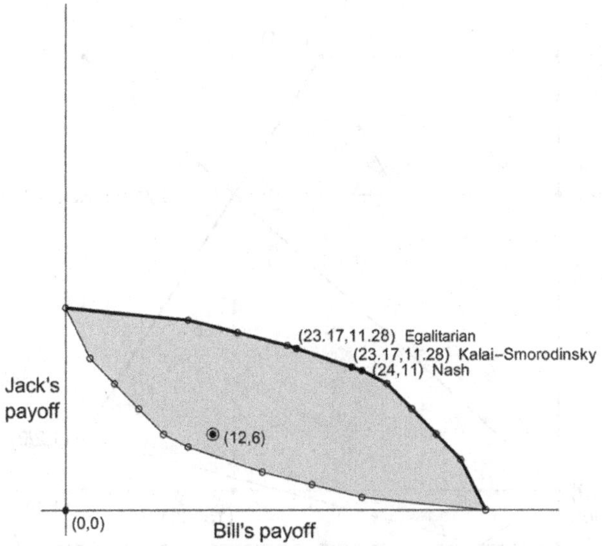

Figure 14 Three axiomatic solutions of the Nash Trading Problem

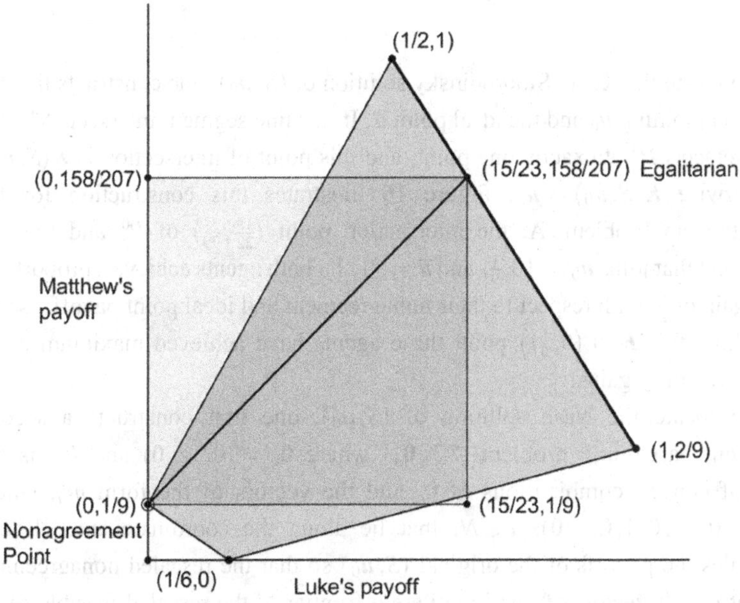

Figure 15 Geometric construction of $E\left(B,\left(0,\frac{1}{9}\right)\right)$

respect to $\boldsymbol{u}_0$, assuming this $0-1$ scale establishes a meaningful standard for comparing their utilities.

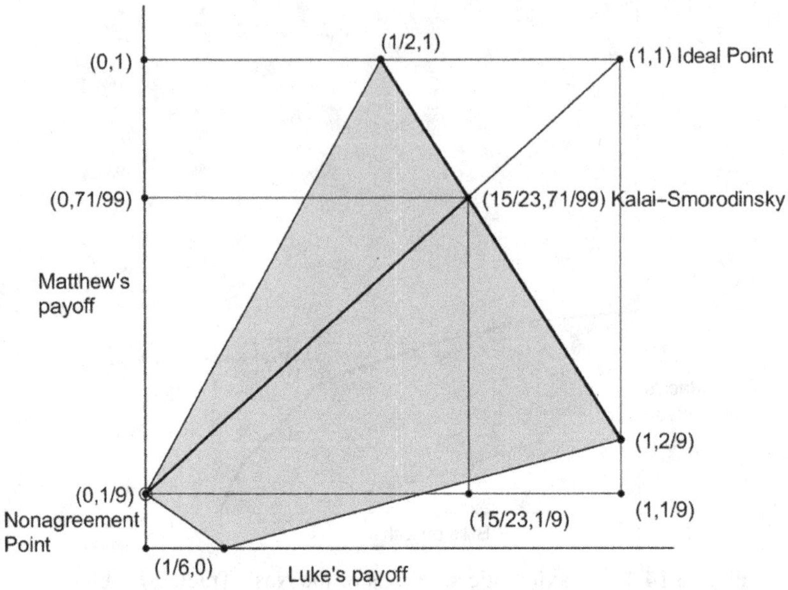

Figure 16 Geometric construction of $\boldsymbol{K}\left(B,\left(0,\frac{1}{9}\right)\right)$

To locate the Kalai–Smorodinsky solution of $(S, \boldsymbol{u}_0)$, one constructs the line segment joining $\boldsymbol{u}_0$ and the ideal point $\overline{\boldsymbol{u}}$. If this line segment intersects S^+ then it intersects P_S^0 at exactly one point, and this point of intersection is $\boldsymbol{K}(S, \boldsymbol{u}_0)$. Otherwise $\boldsymbol{K}(S, \boldsymbol{u}_0) = \boldsymbol{u}_0$. Figure 16 illustrates this construction for the Braithwaite Problem. At the intersection point $\left(\frac{15}{22}, \frac{71}{99}\right)$ of P_B^0 and the line segment that joins $\boldsymbol{u}_0 = \left(0, \frac{1}{9}\right)$ and $\overline{\boldsymbol{u}} = (1, 1)$, both agents achieve a proportionate gain of $\frac{15}{22}$ with respect to their nonagreement and ideal point payoffs, so by reaching this $\boldsymbol{K}\left(B,\left(0,\frac{1}{9}\right)\right)$ point these agents have achieved maximum equal relative utility gains.

To locate the Nash solution of $(S, \boldsymbol{u}_0)$, one first constructs a second n-agent bargaining problem $(U^n, \mathbf{0}_n)$ where $\mathbf{0}_n = (0, \ldots, 0)$ and U^n is the set of convex combinations of $\mathbf{0}_n$ and the vectors of the form $n\boldsymbol{e}_i$, where $\boldsymbol{e}_i = (0, \ldots, 0, 1, 0, \ldots 0)$, $i \in N$, that lie along the coordinate axes. If one rescales the payoffs of the original $(S, \boldsymbol{u}_0)$ so that the rescaled nonagreement point payoff vector is $\mathbf{0}_n$ and the Pareto frontier of the rescaled feasible set S' comes into contact with the Pareto frontier of U^n, then the payoff vectors of the cooperative surplus set S'^+ of the rescaled problem are now embedded in $(U^n, \mathbf{0})$. The Nash solution of the rescaled $(S', \mathbf{0}_n)$ and $(U^n, \mathbf{0}_n)$ coincide at $\mathbf{1}_n$, and the Nash solution of the original $(S, \boldsymbol{u}_0)$ is found by applying the inverse of the rescaling that produced $(S', \mathbf{0}_n)$ to $\mathbf{1}_n$. Figure 17 illustrates the embedding

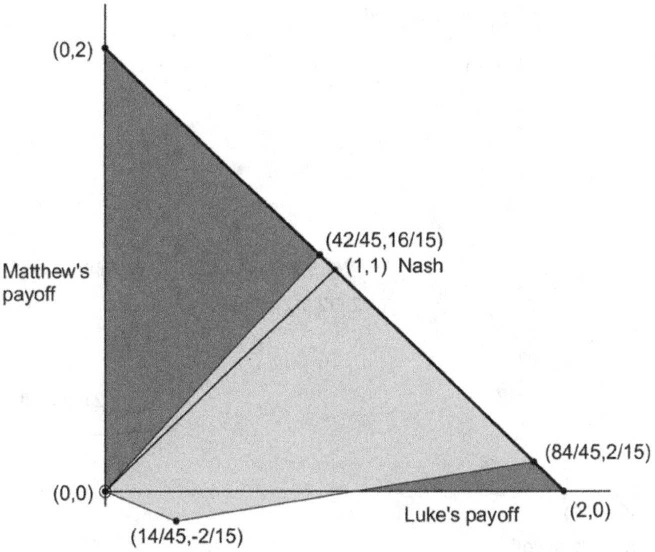

Figure 17 Braithwaite Problem embedded in $(U^2, \mathbf{0}_2)$

part of this construction for the Braithwaite Problem. Here U^2 is the compact triangular region with vertices $(2,0)$, $(0,2)$, and $(0,0)$, depicted as the light shaded region of Figure 17. The payoff vectors of the original $(B, \boldsymbol{u}_0)$ are rescaled according to the transformation $(u_1{}', u_2{}') = \left(\frac{28}{15} \cdot u_1, \frac{18}{15} \cdot u_2 - \frac{2}{15}\right)$ for $(u_1, u_2) \in B$, depicted as the superimposed dark shaded region of Figure 17. The Nash solution of the rescaled Braithwaite Problem $(B', \mathbf{0}_2)$ is $\mathbf{N}(B', \mathbf{0}_2) = (1,1) = \mathbf{1}_2$, which is also the Nash solution of the $(U^2, \mathbf{0}_2)$ triangular problem that contains all of the embedded coöperative surplus payoff vectors of $(B', \mathbf{0}_2)$. By reaching the $\boldsymbol{N}(B', \mathbf{0}_2)$ point these agents have reached the equal division point of an extension of the $(B', \mathbf{0}_2)$ cooperative surplus set B'^{+} that is itself structurally equivalent to a version of the Chocolate Cake Problem where at the nonagreement point each claimant simply gets no cake and where equal division is the "natural" solution. The Nash solution of the original $(B, \boldsymbol{u}_0)$ is obtained by applying the inverse transformation to $\mathbf{1}_2$, yielding $\boldsymbol{N}(B, (0, \frac{1}{9})) = \left(\frac{15}{28} \cdot 1, \frac{15}{18} \cdot 1 + \frac{1}{9}\right) = \left(\frac{15}{28}, \frac{17}{18}\right)$.

The Nash, the Kalai–Smorodinsky, and the Egalitarian solutions all have extensions that are important in applications to social contract theory. For a given $\boldsymbol{u} = (u_1, \ldots, u_n) \in S$ of the n-agent bargaining problem $(S, \boldsymbol{u}_0)$, given $\alpha_i \geq 0$ for each Agent i, the *$\boldsymbol{\alpha}$-weighted Nash product* where $\boldsymbol{\alpha} = (\alpha_1, \ldots, \alpha_n)$ is $(u_1 - u_{01})^{\alpha_1} \cdots (u_n - u_{0n})^{\alpha_n}$, and the *$\boldsymbol{\alpha}$-weighted Nash solution* of this problem

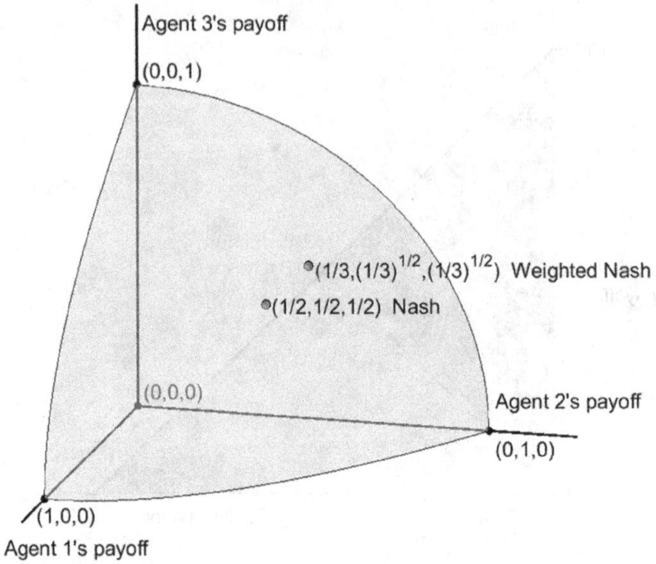

Figure 18 Ordinary and $\boldsymbol{\alpha}$-weighted Nash solutions of $(W, \mathbf{0}_3)$

is the payoff vector $\boldsymbol{N}^{\boldsymbol{\alpha}}(S, \boldsymbol{u}_0) \in S$ that defines the maximum of these products.[24] The ordinary Nash solution is a special case where $\boldsymbol{\alpha} = (1, \ldots, 1)$. Figure 18 illustrates an example of this concept applied to the Wine Division Problem $(W, \mathbf{0}_3)$ of Section 3. The ordinary Nash solution of $(W, \mathbf{0}_3)$ is $\boldsymbol{N}(W, \mathbf{0}_3) = \left(\frac{1}{2}, \frac{1}{2}, \frac{1}{2}\right)$, with the corresponding claim profile $\boldsymbol{x}_N = \left(\frac{1}{2}, \frac{1}{4}, \frac{1}{4}\right)$. For $\boldsymbol{\alpha} = (1, 2, 2)$, the $\boldsymbol{\alpha}$-weighted Nash solution is $\boldsymbol{N}^{\boldsymbol{\alpha}}(W, \mathbf{0}_3) = \left(\frac{1}{3}, \left(\frac{1}{3}\right)^{\frac{1}{2}}, \left(\frac{1}{3}\right)^{\frac{1}{2}}\right)$, with the corresponding claim profile $\boldsymbol{x}_{N^{\alpha}} = \left(\frac{1}{3}, \frac{1}{3}, \frac{1}{3}\right)$. John Harsanyi and Reinhard Selten (1972) introduced the weighted Nash solution concept. One can interpret the possibly different values of the α_i s in an n-agent bargaining problem as reflecting some differential power among the agents relevant to their problem. One possible source of such differential bargaining power is rooted in how agents discount future payoffs, so that α_i reflects Agent i's patience in the bargaining process as an increasing function of Agent i's *discount factor*. So one can view $\boldsymbol{\alpha} = (1, 2, 2)$ applied to the Wine Division Problem as reflecting the assumption that the Frenchman and the Spaniard both have higher discount factors for future payoffs than the German's discount factor, so that the Frenchman and the Spaniard are both more patient than the German and consequently have somewhat greater bargaining power on this dimension than the German.

---

[24] $\boldsymbol{N}^{\boldsymbol{\alpha}}(S, \boldsymbol{u}_0)$ is well defined for reasons similar to the reasons $\boldsymbol{N}(S, \boldsymbol{u}_0)$ is well defined, summarized in note 16.

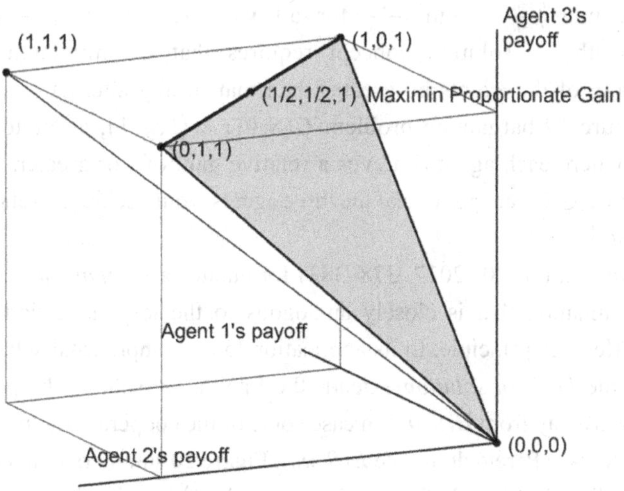

Figure 19 Roth example

David Gauthier (1985, 1986, 136–155) proposes a *Maximin Proportionate Gain* solution that plays a central role in the contractarian moral theory he develops in *Morals by Agreement*.[25] The Maximin Proportionate Gain solution extends the Kalai–Smorodinsky solution by modifying the requirement that the agents achieve equal relative gains from $\boldsymbol{u}_0$. For some bargaining problems with three or more agents where the feasible set is not $\boldsymbol{u}_0$-comprehensive, the Kalai–Smorodinsky solution is intuitively unappealing. Alvin Roth 1979, 105–107) presents a striking example of this phenomenon that is summarized in Figure 19. For this bargaining problem $(S, \boldsymbol{u}_0)$, the feasible set S is the set of convex combinations of $(0, 1, 1)$, $(1, 0, 1)$, and $\mathbf{0}_3$, the nonagreement point is $\boldsymbol{u}_0 = \mathbf{0}_3$, the cooperative surplus set is $S^+ = S - \{\mathbf{0}_3\}$, and the ideal point is $\overline{\boldsymbol{u}} = \mathbf{1}_3$. Since the line segment that joins $\mathbf{0}_3$ and $\mathbf{1}_3$ lies outside S, $\boldsymbol{K}(S, \boldsymbol{u}_0) = \mathbf{0}_3$, the worst possible outcome for all three agents.[26] Gauthier uses Roth's example to motivate his Maximin Proportionate Gain solution concept (1985, 35–36). For a given n-agent bargaining problem $(S, \boldsymbol{u}_0)$ with ideal point $\overline{\boldsymbol{u}}$ and cooperative surplus set S^+, the Maximin Proportionate Gain solution is the payoff vector $\boldsymbol{G}(S, \boldsymbol{u}_0) = (u_1^*, \ldots, u_n^*) \in P_S^+$

[25] This is Gauthier's most recent name for this concept (2013). In *Morals by Agreement* Gauthier calls this concept the *Minimax Relative Concession* solution, because at this solution the maximum concession in terms of the ratio of achieved payoff less nonagreement payoff to ideal payoff less nonagreement payoff is minimized across agents (1986, 137). In both his earlier 1985 essay and in *Morals by Agreement* (1986, 154–155), Gauthier shows that one can define this concept equivalently either in terms of maximin relative gain or maximin relative concession.

[26] In fact, Roth's example summarized here is the three-agent case of a whole family of n-agent bargaining problems Roth presented where $\boldsymbol{K}(S, \boldsymbol{u}_0)$ coincides with $\boldsymbol{u}_0$.

such that $\min_i \frac{u_i^* - u_{0i}}{\bar{u}_i - u_{0i}} > \min_j \frac{v_j - u_{0j}}{\bar{u}_j - u_{0j}}$ for all $\boldsymbol{v} = (v_1, \ldots, v_n) \in S^+ - \{(u_1^*, \ldots, u_n^*)\}$.[27] Gauthier's solution concept requires that the minimum relative gain for any of the agents be greater than that of any alternative outcome. In the Figure 19 bargaining problem $\boldsymbol{G}(S, \boldsymbol{0}_3) = \left(\frac{1}{2}, \frac{1}{2}, 1\right)$, a Pareto optimal outcome where each agent achieves a relative gain of $\frac{1}{2}$ or greater.[28] At any other point $\boldsymbol{v} \in S^+$ at least one of the three agents would achieve a relative gain of less than $\frac{1}{2}$.

Amartya Sen ((1970) 2017, 378–383) formulates a *Leximin rule* for social welfare evaluation that is closely analogous to the lexical version of John Rawls' difference principle. In its application to $\boldsymbol{u}_0$-comprehensive bargaining problems the *Leximin solution* extends the Egalitarian solution by permitting some moves away from $\boldsymbol{E}(S, \boldsymbol{u}_0)$ in case some of the cooperative surplus points of $(S, \boldsymbol{u}_0)$ weakly Pareto dominate $\boldsymbol{E}(S, \boldsymbol{u}_0)$. Figure 20 summarizes a revision of Nash's Trading Problem. In this problem $(R', (12, 6))$, the feasible set of Nash's

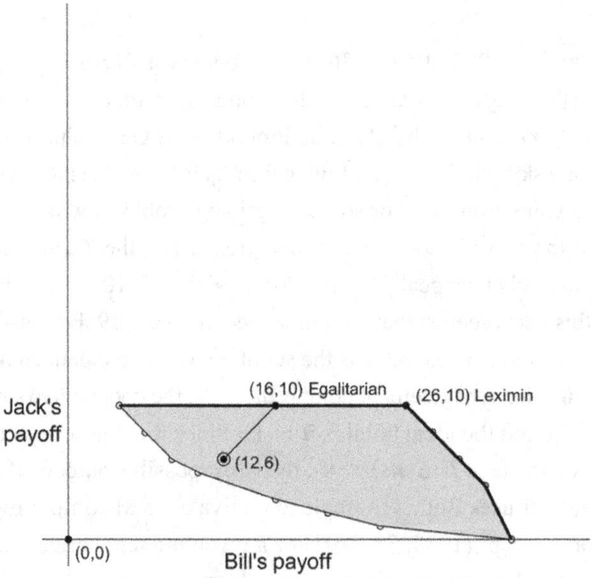

Figure 20 Reduced Nash Trading Problem

[27] Gaertner and Klemisch-Ahlert give this definition of the Maximin Proportionate Gain solution, which corrects Gauthier's original definition by requiring that $\boldsymbol{G}(S, \boldsymbol{u}_0) \in P_S^+$. In his original definition Gauthier did not specify that $\boldsymbol{G}(S, \boldsymbol{u}_0)$ is Pareto optimal, and Gaertner and Klemisch-Ahlert show by example that Gauthier's original definition can fail to identify a unique solution vector in S (1992, 87–88). Gaertner and Klemisch-Ahlert show that the corrected definition identifies a unique $\boldsymbol{G}(S, \boldsymbol{u}_0) \in S$. They also show that for a wide class of bargaining problems a *Lexicographic Kalai–Smorodinsky* solution proposed by Imai (1983) coincides with $\boldsymbol{G}(S, \boldsymbol{u}_0)$ (1992, 88–91, 95–106).

[28] For this bargaining problem $\left(\frac{1}{2}, \frac{1}{2}, 1\right)$ is also the ordinary Nash solution.

original problem is reduced by eliminating all of the trades of the original R where Jack achieves a payoff greater than 10. $(R',(12,6))$ is $(12,6)$-comprehensive, but unlike the original $(R,(12,6))$ is not strictly $(12,6)$-comprehensive. The Egalitarian solution is $\boldsymbol{E}(R',(12,6)) = (16,10)$. One might question the acceptability of $\boldsymbol{E}(R',(12,6))$ as a solution for Bill and Jack in this reduced problem, since all of the points of the form $(16+u_1,10)$ where $0 < u_1 \leq 10$ lie along the undominated boundary $P^0_{R'}$ and all weakly Pareto dominate $\boldsymbol{E}(R',(12,6))$. Jack would fare just as well and Bill would fare better at any of these points than they fare at $\boldsymbol{E}(R',(12,6))$. The Leximin solution allows Jack and Bill to move from $\boldsymbol{E}(R',(12,6))$ to one of these "better" points. One definition of the Leximin solution uses a lexical ordering of all points in $\mathbb{R}^n$: Given $\boldsymbol{x} = (x_1,\ldots,x_n) \in \mathbb{R}^n$, $\gamma(\boldsymbol{x}) = (\gamma_1(\boldsymbol{x}),\ldots,\gamma_n(\boldsymbol{x})) \in \mathbb{R}^n$ is the vector obtained by permuting the coordinates of $\boldsymbol{x}$ in nondecreasing numeric order, so that $\gamma_1(\boldsymbol{x}) \leq \gamma_2(\boldsymbol{x}) \leq \cdots \leq \gamma_n(\boldsymbol{x})$. Then for $\boldsymbol{x},\boldsymbol{y} \in \mathbb{R}^n$, $\boldsymbol{x}$ is *lexicographically greater than* $\boldsymbol{y} \neq \boldsymbol{x}$, $\boldsymbol{x} \succ_L \boldsymbol{y}$, if $\gamma_1(\boldsymbol{x}) > \gamma_1(\boldsymbol{y})$ or if for some $k > 1$, $\gamma_i(\boldsymbol{x}) = \gamma_i(\boldsymbol{y})$ for $1 \leq i < k$ and $\gamma_k(\boldsymbol{x}) > \gamma_k(\boldsymbol{y})$. $\boldsymbol{x}$ is the *lexicographic maximum element* of a set $\Lambda \subset \mathbb{R}^n$ if $\boldsymbol{x} \in \Lambda$ and $\boldsymbol{x} \succ_L \boldsymbol{y}$ for all $\boldsymbol{y} \in \Lambda - \{\boldsymbol{x}\}$. Given the lexical ordering $\succ_L$, the Leximin solution of the $\boldsymbol{u}_0$-comprehensive $(S,\boldsymbol{u}_0)$ is the vector $\boldsymbol{L}(S,\boldsymbol{u}_0) \in S$ that is the lexicographic maximum element of S^+. Many sets in $\mathbb{R}^n$ have no lexicographic maximum element, but the cooperative surplus set of a $\boldsymbol{u}_0$-comprehensive bargaining problem does have such an element,[29] so for $\boldsymbol{u}_0$-comprehensive bargaining problems the Leximin solution is well defined. For the n-agent bargaining problem $(S,\boldsymbol{u}_0)$, the Leximin rule effectively first maximizes the payoff in S^+ of the worst-off agent, then given this constraint maximizes the payoff in S^+ of the second worst-off agent, and so on. So the Leximin rule indeed resembles Rawls' lexical difference principle for assigning shares of primary social goods to representative members of society (1971, 83). In Bill and Jack's reduced bargaining problem of Figure 20, the Leximin solution "corrects" the Egalitarian solution payoff assignment by moving from $\boldsymbol{E}(R',(12,6))$ to the strictly Pareto optimal point $\boldsymbol{L}(R',(12,6)) = (26,10)$ where Jack fares just as well as he fares at $\boldsymbol{E}(R',(12,6))$, and Bill achieves his best possible payoff given Jack's fixed payoff.

5 Rational Choice Justifications of Solutions

The concepts reviewed in Section 4 are only a small subset of the solution concepts game theorists and philosophers have considered since Nash and

29 This follows because the union $S^+ \cup \{\boldsymbol{u}_0\}$ of the cooperative surplus set and the nonagreement point is a convex and compact set.

Raiffa published their foundational works. Is any solution concept the *right* one? Perhaps the best answer to this question is: It depends. More specifically, which solution concept is the "right" concept turns upon what one requires of and what one is willing to leave as optional for a solution concept, which might vary according to context. Many solution concepts satisfy some apparently desirable properties, but none generally satisfies all of a surprisingly short list of such properties.

One tradition explores which outcome rational agents equipped with full knowledge of the structure of the bargaining problem would select. Nash launched this rational choice tradition. Nash was responding to von Neumann and Morgenstern, who thought that rational agents engaged in a problem involving bargaining might arrive at any member of what they identified as a *stable set* of outcomes.[30] Nash maintained that the bargaining problem should have a determinate solution that rational agents will identify and follow. Nash also argued that in principle one can analyze any bargaining problem either *axiomatically* by considering which utility allocations satisfy certain formal desiderata, or *strategically* by exploring how agents might arrive at some outcome resulting from a sequence of moves reflecting some actual bargaining process. One can view the axiomatic approach as analyzing the bargaining problem from the perspective of an ideally rational arbiter who knows all of the structural properties of the feasible set. Such an arbiter is in a position analogous to Adam Smith's impartial spectator ((1759) 1982, 129, 135), who in this context chooses a solution of the agents' bargaining problem on their behalf. The axiomatic approach is also Hobbesian in spirit, since Hobbes claims distributive justice is the justice of an arbiter ((1651) 1994, 15:14). Indeed, axiomatic solution concepts are sometimes called *arbitration schemes*. The axioms that summarize the formal desiderata of a solution presumably specify any relevant constraints on the arbiter's rational choice in this context. One can view the strategic approach as analyzing the bargaining problem from the perspectives of the agents themselves should they try to arrive at a solution via the interaction of their individually rational moves. The strategic approach typically starts from the assumptions that the agents in the bargaining problem are (Bayesian) rational, so that each consistently applies the Bayesian standard of choosing actions that maximize their expected utility, and that these agents have common knowledge of their rationality and their payoffs at the different alternative feasible set points.

[30] Luce and Raiffa give an overview of von Neumann and Morgenstern's analysis of bargaining (1957, 115–118). Von Neumann and Morgenstern's original discussions of bargaining are somewhat scattered in various sections of *Theory of Games and Economic Behavior* ((1944) 2004).

Nash himself presented an alternate way of viewing the axiomatic and the strategic analyses of the bargaining problem via the distinction between cooperative and noncooperative game theory. In his essay "Non-Cooperative Games," Nash argued that one can regard any cooperative game as the output of a larger noncooperative game containing the moves available to the agents that enable them to form the coalitions and the binding agreements of this cooperative game (1951a, 295). The general project of generating cooperative game theory from noncooperative game theory, one that game theorists have yet to complete, is known as the *Nash program*. One can view strategic analyses of the bargaining problem as a special case of the Nash program. From the axiomatic perspective a bargaining problem is a cooperative game where all the agents together act as a coalition, whose members all must follow a given outcome as their solution or else be left at their nonagreement point payoffs. And from the strategic perspective a bargaining problem is embedded in a larger non-cooperative game, such that the agents can reach a given outcome as their solution by following some strategy profile of this larger game. Nash thought that the axiomatic and strategic bargaining approaches should yield complementary results (1953, 129). Each of these two approaches has motivated a large body of research that continues to grow. Game theorists often present large parts of axiomatic or strategic bargaining theory independently of the other theory. Yet, just as Nash believed, aspects of each theory help to motivate and to illuminate the other.

5.1 Nash's Axiomatic and Strategic Analyses

Nash's own arguments in defense of his solution concept remain great illustrations of the axiomatic and strategic approaches. In his first essay on the bargaining problem, Nash (1950) argued that a bargaining problem should have a unique solution, and that this solution should satisfy several properties he states precisely as axioms. One of these axioms can be stated in terms of an *affine transformation* of payoffs, and another incorporates a notion of *symmetry*. An affine transformation is a function $T : \mathbb{R}^n \rightarrow \mathbb{R}^n$ where $T(u_1, \ldots, u_n) = (a_1u_1 + b_1, \ldots, a_nu_n + b_n)$ with $a_i > 0$ and $b_i \in \mathbb{R}$ for each $i \in N$. When the payoffs of S are von Neumann–Morgenstern utilities, they are unique only up to a choice of scales, and in this case an affine transformation applied to S defines a set of rescalings of these payoffs. A bargaining problem $(S, \boldsymbol{u}_0)$ is *symmetric* when for any permutation $\pi(\boldsymbol{u})$ of the components of a point $\boldsymbol{u} = (u_1, \ldots, u_n) \in S$ we have $\pi(\boldsymbol{u}) \in S$ and $\pi(\boldsymbol{u}_0) = \boldsymbol{u}_0$. Informally, $(S, \boldsymbol{u}_0)$ is symmetric when the feasible set S and the nonagreement point $\boldsymbol{u}_0$

are unchanged if one interchanges the positions of the agents. I now state a set of axioms that are equivalent to the axioms Nash proposed for a solution $\boldsymbol{f}(S, \boldsymbol{u}_0) = (u_1^*, \ldots, u_n^*) \in S$ of an essential bargaining problem:[31]

Weak Mutual Advantage: $\boldsymbol{f}(S, \boldsymbol{u}_0) \geqq \mathrm{u}_0$.
Pareto Optimality: $\boldsymbol{f}(S, \boldsymbol{u}_0) \in P_S^+$.
Scale Invariance: If one rescales the payoffs of $(S, \boldsymbol{u}_0)$ according to the affine transformation $T : \mathbb{R}^n \rightarrow \mathbb{R}^n$, then $\boldsymbol{f}(T(S), T(\boldsymbol{u}_0)) = T(\boldsymbol{f}(S, \boldsymbol{u}_0))$.
Symmetry: If $(S, \boldsymbol{u}_0)$ is symmetric, then $u_1^* = \cdots = u_n^*$.
Contraction Consistency: If $S_0 \subset S_1$ and $\boldsymbol{f}(S_1, \boldsymbol{u}_0) \in S_0$, then $\boldsymbol{f}(S_0, \boldsymbol{u}_0) = \boldsymbol{f}(S_1, \boldsymbol{u}_0)$.[32]

What do Nash's axioms say about a solution $\boldsymbol{f}(S, \boldsymbol{u}_0)$? Weak Mutual Advantage says that at $\boldsymbol{f}(S, \boldsymbol{u}_0)$ each Agent i fares at least as well as they fare at the nonagreement point $\boldsymbol{u}_0$.[33] Weak Mutual Advantage limits the candidates for a solution to the nonagreement point $\boldsymbol{u}_0$ itself and payoff vectors in the cooperative surplus set S^+. By definition, Pareto Optimality says no payoff vector of S weakly dominates $\boldsymbol{f}(S, \boldsymbol{u}_0)$. Pareto Optimality reflects the idea that nothing should go to waste, for if any $\boldsymbol{v} \in S$ did weakly dominate $\boldsymbol{f}(S, \boldsymbol{u}_0)$ then there would be "unused utility" at $\boldsymbol{f}(S, \boldsymbol{u}_0)$, in the sense that by moving from $\boldsymbol{f}(S, \boldsymbol{u}_0)$ to $\boldsymbol{v}$ some would reach higher payoffs without any having to accept lower payoffs. Scale Invariance says that the utilities of S are not comparable across agents, reflecting Nash's express view that the payoffs of a bargaining problem are von Neumann–Morgenstern utilities. When the payoffs of a bargaining problem are defined in terms of some distributed goods like Claudia and Laura's chocolate cake or the wine of Hume's example, Scale Invariance ensures that one cannot change anyone's assigned share of the goods by choosing new scales for any of the agents' payoffs in S. Symmetry says that if one has exactly the same nonagreement point and feasible set after one interchanges the positions of the payoff vector components of S, then the solution must assign the same payoff to all the agents. Contraction Consistency says that if $\boldsymbol{f}(S_1, \boldsymbol{u}_0)$ is the solution of a bargaining problem $(S_1, \boldsymbol{u}_0)$ and $\boldsymbol{f}(S_1, \boldsymbol{u}_0)$ is still in the feasible set S_0 of a reduction $(S_0, \boldsymbol{u}_0)$ of $(S_1, \boldsymbol{u}_0)$, then $\boldsymbol{f}(S_1, \boldsymbol{u}_0)$ is also the solution of the

31 The axioms I state here are based primarily upon the alternate restatements of Nash's original axioms in Luce and Raiffa 1957, 126–127) for the two-agent case and Thomson and Lensberg 1989, 13–14) for the n-agent case. Nash (1950) states his axioms somewhat differently and only for the two-agent case.

32 This axiom is also known as *Independence of Irrelevant Alternatives*. However, this axiom of bargaining solutions should not be confused with the constraint on social preferences that Kenneth Arrow gives the same name in his foundational work on constructing a social preference from individual preferences ((1951) 2012, 26), a name that has become standard in the social choice literature.

33 This axiom is also known as *Weak Individual Rationality* since it says that at $\boldsymbol{f}(S, \boldsymbol{u}_0)$ no agent has to make the effectively irrational choice of accepting a lower payoff than they would receive at $\boldsymbol{u}_0$.

reduction $(S_0, \boldsymbol{u}_0)$. Equivalently, Contraction Consistency says that if $S_1 \subset S_2$ so that $(S_2, \boldsymbol{u}_0)$ is an expansion of $(S_1, \boldsymbol{u}_0)$, then a solution of the expansion $(S_2, \boldsymbol{u}_0)$ must be either $\boldsymbol{f}(S_1, \boldsymbol{u}_0)$ itself or some point $\boldsymbol{v}^* \in S_2 - S_1$ outside the original S_1. Contraction Consistency, which might alternately be called *Contraction-Expansion Consistency*, ensures that a "losing" payoff vector of a bargaining problem cannot be "promoted" into being a solution either by enlarging the feasible set or by reducing this set in such a way that the original "winner" remains.

Nash showed that if $(S, \boldsymbol{u}_0)$ is an essential bargaining problem, then for any $\boldsymbol{f}(S, \boldsymbol{u}_0)$ that satisfies his axioms $\boldsymbol{f}(S, \boldsymbol{u}_0) = \boldsymbol{N}(S, \boldsymbol{u}_0)$, the Nash solution.[34] The monumental insight of Nash's 1950 essay was that only a few formal properties could characterize a *function* over the domain of essential bargaining problems that selects a unique solution for each such problem. This is why using functional notation to refer to the payoff vector of Nash's solution makes sense. When Nash's axioms are satisfied, $\boldsymbol{f}$ is a well-defined function that assigns to each essential $(S, \boldsymbol{u}_0)$ a unique payoff vector $\boldsymbol{f}(S, \boldsymbol{u}_0) \in S$ as its solution.[35] In fact, given that Nash effectively limits the domain of the function $\boldsymbol{f}$ to essential bargaining problems, the Nash solution satisfies another general property summarized by the axiom

Strong Mutual Advantage: $\boldsymbol{f}(S, \boldsymbol{u}_0) > \boldsymbol{u}_0$.

Strong Mutual Advantage, a strengthening of Weak Mutual Advantage, says that at $\boldsymbol{f}(S, \boldsymbol{u}_0)$ all claimants fare strictly better than they fair at the nonagreement point $\boldsymbol{u}_0$. Alvin Roth (1977) showed that the Nash solution is the unique solution that satisfies Scale Invariance, Symmetry, Contraction Consistency, and Strong Mutual Advantage. Roth's result illustrates how a solution concept can be characterized by more than one set of axioms.

In his second essay on bargaining Nash (1953) presented a model of a strategic process for selecting a solution of a two-agent bargaining problem with feasible set S. Nash's model has two stages. At the first stage, each Agent i independently chooses a strategy s_i, and at the strategy profile $\boldsymbol{s} = (s_1, s_2)$ they achieve the payoff vector $\boldsymbol{u}(\boldsymbol{s}) = \left(u_1(\boldsymbol{s}), u_2(\boldsymbol{s})\right)$ where $u_i(\boldsymbol{s})$ is a payoff available to Agent i in S. At the second stage, they engage in a Nash demand game where each Agent i's pure strategies are the payoffs available to them in S and $\boldsymbol{u}(\boldsymbol{s})$ is their nonagreement payoff vector. Each independently claims a payoff v_i

34 Nash gave a somewhat informal proof of this result for the two-agent case (1950, 159). Many authors have published formal proofs of Nash's result extended to the n-agent case based upon Nash's original argument. Thomson and Lensberg (1989, 15) give an elegant formal proof.

35 In fact, my use of "$\boldsymbol{f}(S, \boldsymbol{u}_0)$" to refer to the value of the function $\boldsymbol{f}$ at the argument $(S, \boldsymbol{u}_0)$ is a slight abuse of the usual functional notation $\boldsymbol{f}\left((S, \boldsymbol{u}_0)\right)$, an abuse meant to make the notation less clumsy.

available in S and receives v_i given compatible claims and $u_i(\boldsymbol{s})$ otherwise, because given incompatible claims each Agent i must follow s_i. Nash also shows that given his construction at least one profile $\boldsymbol{s}^* = (s_1^*, s_2^*)$ of first stage strategies is in equilibrium – that is, each of these strategies is a best reply to the other. And Nash concludes that the two agents, being rational, will identify and if necessary follow such an $\boldsymbol{s}^*$. Nash's model effectively embeds the bargaining problem $(S, \boldsymbol{u}(\boldsymbol{s}^*))$ in the noncooperative game defined by the agents' possible moves at the two stages. At the first stage they determine their nonagreement point by choosing their *threat strategies* s_1^* and s_2^*. At the second stage they engage in a subgame that is their Nash demand game with nonagreement point payoffs determined by their threat strategies. After they have chosen s_1^* and s_2^*, these threat strategies are *credible* in that in the event they issue incompatible claims at the second stage, neither Agent i can do better by deviating from s_i^* given that they expect their counterpart Agent j to follow s_j^*. As discussed in Section 3, Nash's construction extends S into a $\boldsymbol{u}(\boldsymbol{s}^*)$-comprehensive set if the original S is not already $\boldsymbol{u}(\boldsymbol{s}^*)$-comprehensive. Also as discussed in Section 3, each point $\boldsymbol{u} \geq \boldsymbol{u}(\boldsymbol{s}^*)$ in the set of compatible claims vectors that is not weakly dominated by some other point in this set is a Nash equilibrium of the demand game. To select one of these equilibria, Nash next considers a *smoothed demand game* that approximates the original demand game by making demand profiles that lie outside S incompatible only with a probability defined by a distribution function. Nash shows that the smoothed demand game has a unique Nash equilibrium and that if the "smoothing effect" of the distribution function approaches zero, this equilibrium converges in the limit to the Nash solution $\boldsymbol{N}(S, \boldsymbol{u}(\boldsymbol{s}^*))$ of the original demand game. Nash then presents axioms that characterize the solution of his strategic model, several of which resemble the axioms of his original 1950 axiomatic solution. But Nash includes two new axioms reflecting the fact that in the strategic model the agents must choose their nonagreement point. One of these new axioms requires that if either Agent i's strategy set is reduced while the other Agent j's strategy set remains the same, Agent i's prospects are not improved. The other requires that if either Agent i is limited to a single strategy, then there exists a way to limit the other Agent j to a single strategy without raising Agent i's prospects.

Nash's original work set the agenda for rational choice analyses of the bargaining problem, but the specifics of his axiomatic and strategic characterizations of his own solution came under challenge quite soon after he published his 1950 and 1953 essays. Luce and Raiffa gave an early critique of Nash's strategic model, arguing that while Nash's use of the smoothed demand game to select a single equilibrium is mathematically ingenious, it is simply unclear why Nash's smoothed game is relevant to the agents' problem of trying to reach a mutually satisfactory solution. Luce and Raiffa also agree that Nash's axiomatic characterization of his strategic model solution complements his strategic analysis very well, but are unsure one would have any reason to adopt the new axioms Nash introduces in his 1953 essay, other than that they happen to form part of this complementarity (1957, 141–143). Luce and Raiffa further argue using claimed counterexamples that in his original 1950 axiomatic analysis, each of the Scale Invariance, Symmetry, and Contraction Consistency axioms might at least in certain circumstances fail to reflect the actual situation of agents engaged in a bargaining scenario (1957, 131–134). Criticisms of Nash's original work such as these have helped to motivate alternative axiomatic and strategic defenses of both the Nash solution and other solution concepts. Indeed, theorists have developed a great many distinct axiomatic and strategic characterizations of solution concepts, many of which differ from each other only in subtle ways. In the remainder of this section I will discuss a few of what I take to be the most interesting and important of these characterizations.[36]

5.2 Alternative Axioms Related to Feasible Set Changes

Many of the alternative proposed axioms for bargaining problems track how a solution concept responds to changes in the feasible set S. In this section I will discuss axioms related to bargaining problems of the sort where the feasible set S may change while the set of agents engaged remains the same. Informally, a change in S might reflect some change in the background conditions that define the relevant bargaining problem. For example, S might possibly expand or contract if the payoffs of this set are generated by a division problem where the supply of the goods at stake increases or decreases.

As discussed above, Nash's Contraction Consistency axiom specifies one such stability property with respect to certain contractions and expansions of S: If the original solution $\boldsymbol{f}(S, \boldsymbol{u}_0)$ remains after a contraction or expansion of S, then the solution of the new feasible set S' is either still $\boldsymbol{f}(S, \boldsymbol{u}_0)$ or one of any new payoff vectors that might be present in the new S'. Contraction Consistency is an application of a principle that Amartya Sen in *Collective Choice and Social Welfare* calls *Property* α: If some element $x \in A_2 \subset A_1$ is ranked best in A_1, then x is ranked best in the set A_2 that is a subset of A_1. Sen maintains that Property α is a basic requirement of rational choice ((1970) 2017, 63).

[36] Roth (1979), Thomson and Lensberg (1989), and Peters (1992) give far more extensive presentations of axiomatic justifications of solution concepts. Peters (1992, Chapter 9) also gives a fine overview of several important strategic models applied to the bargaining problem and how these strategic models are related to axiomatic arguments for solution concepts.

A number of other well-known axioms are related to contractions and expansions of S. I state three of these axioms here. First, there is

Restricted Monotonicity: If $(S_1, \boldsymbol{u}_0)$ and $(S_2, \boldsymbol{u}_0)$ have the same ideal point $\overline{\boldsymbol{u}}$ and if $S_1 \subset S_2$, then $\boldsymbol{f}(S_1, \boldsymbol{u}_0) \leqq \boldsymbol{f}(S_2, \boldsymbol{u}_0)$.

Restricted Monotonicity says that if the feasible set is enlarged without changing the ideal point, then each claimant fares at least as well at the resulting new solution as they fare at the old solution – that is, the new solution is either the same as or weakly dominates the old solution. A stronger axiom is

Strong Monotonicity: If $S_1 \subset S_2$, then $\boldsymbol{f}(S_1, \boldsymbol{u}_0) \leqq \boldsymbol{f}(S_2, \boldsymbol{u}_0)$.

Strong Monotonicity says that if the feasible set is enlarged without restriction, then each claimant fares at least as well at the resulting new solution as they fare at the old solution. Strong Monotonicity is closely related to

Decomposability: If $S_1 \subset S_2$, then $\boldsymbol{f}(S_2, \boldsymbol{u}_0) = \boldsymbol{f}(S_2, \boldsymbol{f}(S_1, \boldsymbol{u}_0))$.

Decomposability says that for an expansion $(S_2, \boldsymbol{u}_0)$ of $(S_1, \boldsymbol{u}_0)$, the result is the same whether one applies the solution function $\boldsymbol{f}$ to the larger problem $(S_2, \boldsymbol{u}_0)$ all at once or in stages, first to $(S_1, \boldsymbol{u}_0)$ with $\boldsymbol{u}_0$ as nonagreement point and then again to $(S_2, \boldsymbol{f}(S_1, \boldsymbol{u}_0))$ with $\boldsymbol{f}(S_1, \boldsymbol{u}_0)$ as nonagreement point. Put another way, Decomposability says the solution is stable with respect to this systematic movement of the nonagreement point.

Axioms such as Nash's original axioms and those discussed in the previous paragraph summarize properties one presumably would want any solution of a bargaining problem to satisfy. For example, Strong Monotonicity is meant to capture the informal idea that if the available "pie" becomes larger, then every claimant will end up at least as well off as they would have ended up given the original smaller "pie." Who could object to requiring Strong Monotonicity? As a matter of fact, with the possible exception of Weak Mutual Advantage, each of the allegedly innocent-looking axioms discussed above has been questioned. One can present examples that some take to cast doubt on a given axiom together with solutions that require this axiom. And no solution concept will generally satisfy all of even a short list of axioms. To illustrate, suppose Nash's original axioms are augmented by adding Restricted Monotonicity, a weaker axiom than the "unobjectionable" Strong Monotonicity. Figure 21 depicts a pair of bargaining problems $(S_0, \boldsymbol{0}_2)$ and $(S_1, \boldsymbol{0}_2)$ where $S_0 \subset S_1$ so that $(S_0, \boldsymbol{0}_2)$ is embedded in $(S_1, \boldsymbol{u}_0)$. One can regard $(S_1, \boldsymbol{0}_2)$ as an expansion of $(S_0, \boldsymbol{u}_0)$ or, conversely, $(S_0, \boldsymbol{0}_2)$ as a contraction of $(S_1, \boldsymbol{0}_2)$. In both bargaining problems the nonagreement point is $\boldsymbol{0}_2$ and the ideal point is $\overline{\boldsymbol{u}} = (1, 1)$. In the $(S_0, \boldsymbol{u}_0)$ of Figure 21 the Nash, Egalitarian, and Kalai–Smorodinsky solutions all coincide

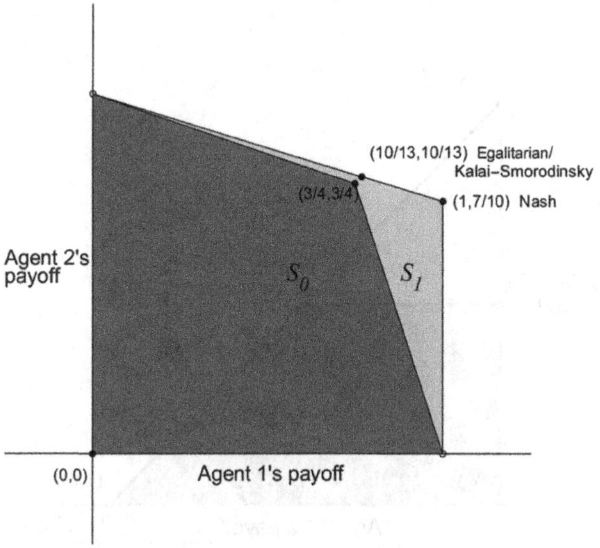

Figure 21 Failure of Restricted Monotonicity example

at $\boldsymbol{u}^* = \left(\frac{3}{4}, \frac{3}{4}\right)$. In the $(S_1, \mathbf{0}_2)$ expansion, the Egalitarian and the Kalai–Smorodinsky solutions coincide at $\boldsymbol{E}(S_1, \mathbf{0}_2) = \boldsymbol{K}(S_1, \mathbf{0}_2) = \left(\frac{10}{13}, \frac{10}{13}\right)$, but the Nash solution is $\boldsymbol{N}(S_1, \mathbf{0}_2) = \left(1, \frac{7}{10}\right)$.[37] So by reapplying the Nash solution in the expanded $(S_1, \mathbf{0}_2)$, Agent 2 fares worse than in the original $(S_0, \mathbf{0}_2)$. The Nash solution does not satisfy Restricted Monotonicity.

At first blush one might conclude that the Figure 21 example shows that one should set aside the Nash solution and adopt some other solution concept, such as the Kalai–Smorodinsky solution or the Egalitarian solution, both of which satisfy Restricted Monotonicity. But they do not satisfy all of Nash's axioms. Figure 22 depicts another pair of bargaining problems where $(S_0, \mathbf{0}_2)$ is embedded in $(S_1, \mathbf{0}_2)$.[38] One can think of the larger $(S_1, \mathbf{0}_2)$ as a version of the Chocolate Cake Problem with a $\mathbf{0}_2$ nonagreement point, and the smaller $(S_0, \mathbf{0}_2)$ as a variant of the Chocolate Cake Problem where Agent 2 is fully sated with half the cake. In $(S_0, \mathbf{0}_2)$, the Kalai–Smorodinsky solution is $\boldsymbol{K}(S_0, \mathbf{0}_2) = \left(\frac{2}{3}, \frac{1}{3}\right)$. This follows from the fact that $\overline{\boldsymbol{u}} = \left(1, \frac{1}{2}\right)$ is the ideal point of $(S_0, \mathbf{0}_2)$. In the larger $(S_1, \mathbf{0}_2)$, the Kalai–Smorodinsky, Nash, and Egalitarian solutions coincide at $\boldsymbol{K}(S_1, \mathbf{0}_2) = \boldsymbol{N}(S_1, \mathbf{0}_2) = \boldsymbol{E}(S_1, \mathbf{0}_2) = \left(\frac{1}{2}, \frac{1}{2}\right)$. When the Kalai–Smorodinsky solution is reapplied to the expansion $(S_1, \mathbf{0}_2)$, Agent 2 gains and Agent 1 loses. This occurs because $\overline{\boldsymbol{u}} = (1, 1)$ is the new ideal point of the expanded $(S_1, \mathbf{0}_2)$.

---

37 This example is due to Kalai and Smorodinsky (1975, 515).

38 This example with a different payoff scaling is due to Luce and Raiffa (1957, 133).

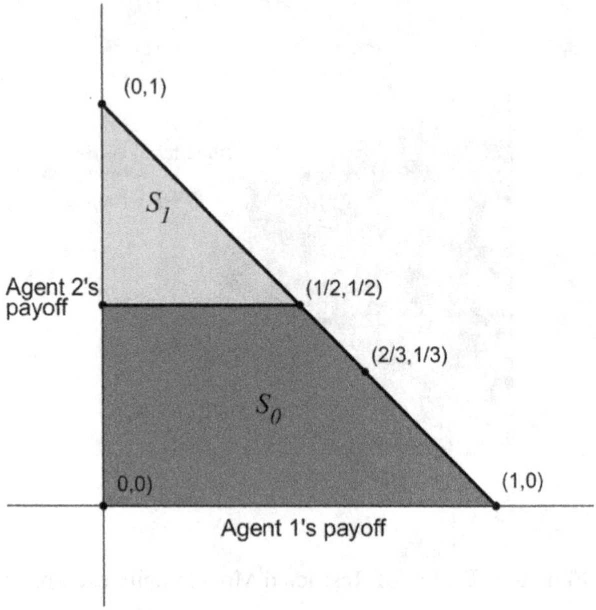

Figure 22 Failure of Contraction Consistency example

This example shows that the Kalai–Smorodinsky solution is not Strongly Monotone. And $\boldsymbol{K}(S_1, \boldsymbol{u}_0) = \left(\frac{1}{2}, \frac{1}{2}\right) \in S_0$, the feasible set of the smaller $(S_0, \mathbf{0}_2)$. The Kalai–Smorodinsky solution "promotes" $\left(\frac{1}{2}, \frac{1}{2}\right) \neq \boldsymbol{K}(S_0, \mathbf{0}_2)$ into the solution of the larger $(S_1, \mathbf{0}_2)$, showing that the Kalai–Smorodinsky solution is not Contraction Consistent. In fact, the Kalai–Smorodinsky solution is characterized by replacing in the set of Nash's axioms Contraction Consistency with Restricted Monotonicity. But must one then give up either Contraction Consistency or Restricted Monotonicity? And if so, which axiom should one give up?

The Egalitarian solution is both Contraction Consistent and Strongly Monotone and therefore also Restricted Monotone. The Egalitarian solution is also Decomposable. But the Egalitarian solution fails to satisfy the Scale Invariance and Pareto Optimality axioms on Nash's list. Scale Invariance fails because the Egalitarian solution assumes the agents have interpersonally comparable utilities calibrated to the same scale, so that transformations of the feasible set can change how much utility some agents can achieve as compared to how much the others can achieve. To illustrate, Figure 23 summarizes another pair of bargaining problems $(S_1, \mathbf{0}_2)$ and $\left(\mathrm{T}(S_2), \mathrm{T}(\mathbf{0}_2)\right)$. Here $(S_1, \boldsymbol{u}_0)$ is the Chocolate Cake Problem of Figure 22, and $\left(\mathrm{T}(S_1), \mathrm{T}(\mathbf{0}_2)\right) = \left(\mathrm{T}(S_1), \mathbf{0}_2\right)$ is an affine transformation of this problem where T multiplies Agent 1's payoffs in S_1

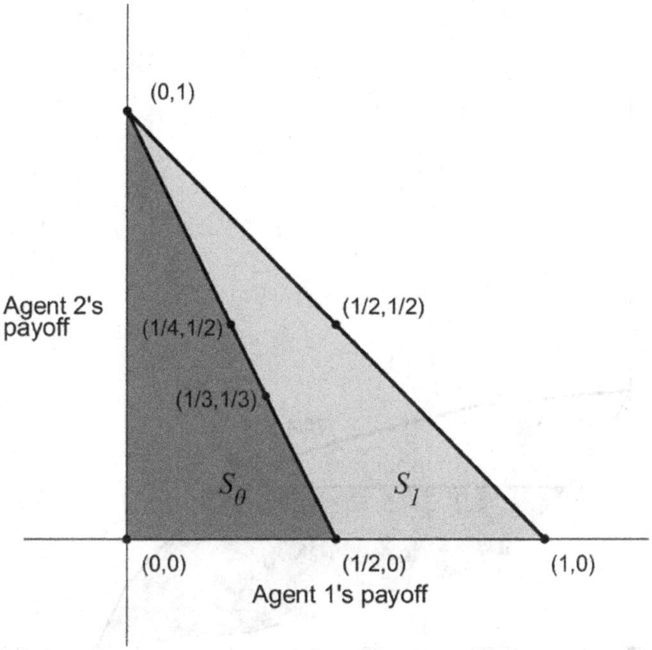

Figure 23 Failure of Scale Invariance example

by a factor of $\frac{1}{2}$. $\boldsymbol{E}(\mathrm{T}(S_1), \mathbf{0}_2) = \left(\frac{1}{3}, \frac{1}{3}\right)$, but $\mathrm{T}\left(\boldsymbol{E}(S_1, \mathbf{0}_2)\right) = \left(\frac{1}{2} \cdot \frac{1}{2}, \frac{1}{2}\right) = \left(\frac{1}{4}, \frac{1}{2}\right)$, so the Egalitarian solution fails Scale Invariance. If one interprets $(S_1, \boldsymbol{u}_0)$ as sharing a Chocolate Cake, then for $(S_1, \mathbf{0}_2)$ at $\boldsymbol{E}(S_1, \boldsymbol{u}_0)$ the corresponding claims profile is $\boldsymbol{x}_{\boldsymbol{E}(S_1, \boldsymbol{u}_0)} = \left(\frac{1}{2}, \frac{1}{2}\right)$. Since utilities are assumed to be interpersonally comparable and S_1 is symmetric, in order to achieve equal utility gains the claimants must receive equal shares of cake. But for $\left(\mathrm{T}(S_2), \mathbf{0}_2\right)$ at $\boldsymbol{E}\left(\mathrm{T}(S_1), \mathbf{0}_2\right)$ the corresponding claims profile is $\boldsymbol{x}_{\boldsymbol{E}(\mathrm{T}(S_1), \mathbf{0}_2)} = \left(\frac{2}{3}, \frac{1}{3}\right)$. Given the assumption that utilities are interpersonally comparable, the transformation T in this context reduces the utility Agent 1 receives from any amount of cake in $(S_1, \mathbf{0}_2)$ by a factor of $\frac{1}{2}$, so that in order to achieve equal utility gains in $\left(\mathrm{T}(S_2), \mathbf{0}_2\right)$ Agent 1 must receive twice as much cake as Agent 2.

Figure 24 depicts both the feasible sets of Figure 2 and Figure 20, where $S_1 = R$ is the feasible set of Nash's Trading Problem first depicted in Figure 2 and $S_0 = R'$ is the reduction of R given in Figure 20. The Figure 20 example shows that the Egalitarian solution does not satisfy Pareto Optimality. $\boldsymbol{E}\left(S_0, (12, 6)\right) = (16, 10)$ is weakly dominated by a set of points in S_0. For

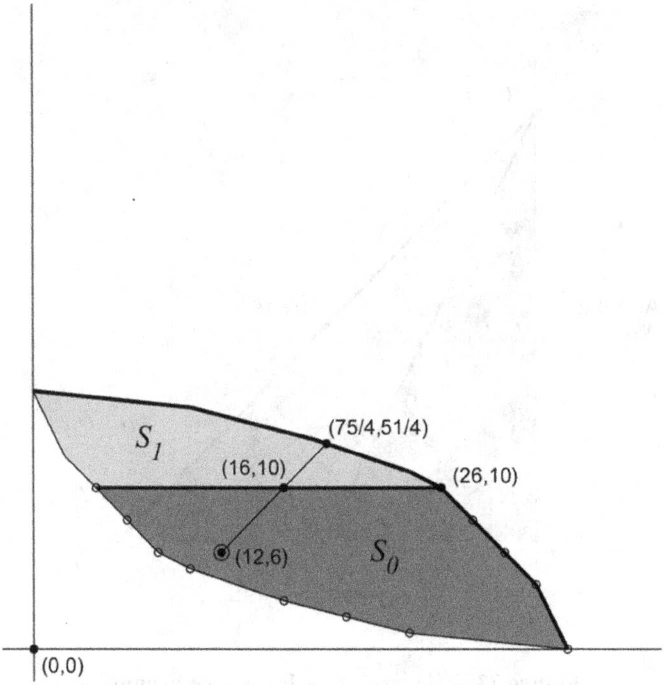

Figure 24 Failure of Pareto Optimality example

$\boldsymbol{u}_0$-comprehensive feasible sets the Egalitarian solution satisfies a weaker property summarized by the axiom

Weak Pareto Optimality: $\boldsymbol{f}(S, \boldsymbol{u}_0) \in P_S^0$.[39]

Weak Pareto Optimality says that no $\boldsymbol{u} \in S$ strictly dominates the solution $f(S, \boldsymbol{u}_0)$ – that is, at no other $\boldsymbol{u} \in S$ can all claimants fare strictly better than they fare at $f(S, \boldsymbol{u}_0)$. As illustrated in Section 4, one can recover Pareto Optimality by adopting the Leximin solution that extends the Egalitarian solution. The Leximin solution of this $(S_0, (12, 6))$ is the Pareto optimal point $\boldsymbol{L}(S_0,(12,6)) = (26, 10)$. In the $(S_1, (12, 6))$ expansion, the Egalitarian and Leximin solutions coincide at $\boldsymbol{E}(S_1,(12,6)) = \boldsymbol{L}(S_1,(12,6)) = (18.75, 12.75)$. $(18.75, 12.75)$ is strictly Pareto optimal, but if they follow the Leximin solution then Bill fares worse in the expansion $(S_1,(12,6))$ than at $(S_0,(12,6))$. The Leximin solution satisfies Pareto Optimality, but this example shows the Leximin solution is not

[39] For $n = 2$ the Kalai–Smorodinsky solution is Pareto optimal, but for $n \geq 3$ and for $\boldsymbol{u}_0$-comprehensive feasible sets the Kalai–Smorodinsky solution is only weakly Pareto optimal.

Monotone. Moving from the Egalitarian solution to the more general Leximin solution secures Pareto Optimality by sacrificing Monotonicity![40]

As the above examples illustrate, there are interesting and important trade-offs of formal desiderata one must make in choosing an axiomatic solution concept. Table 2 summarizes which of the axioms discussed in this section the solution concepts discussed in this Element satisfy over the domain of $\boldsymbol{u}_0$-comprehensive sets.

5.3 Alternative Axioms Related to Adding and Removing Agents

When the number of agents in a bargaining problem can vary, it becomes possible to apply a solution concept to each of a family of specific bargaining problems obtained by adding or removing agents. Some solution concepts can in some sense be stable with respect to such additions or removals. In order to precisely define some of the associated stability properties, some additional notations are needed. Given a vector $\boldsymbol{x} = (x_1, \ldots, x_n)$ and a subset $M \subset N$ where $M = \{i_1, \ldots, i_m : 1 \leq i_1 < \ldots < i_n \leq n\}$, the vector $\boldsymbol{x}_M = (x_{i_1}, \ldots, x_{i_m})$ is the *M-restriction of* $\boldsymbol{x}$. When one starts with the n-dimensional vector $\boldsymbol{x}$, $\boldsymbol{x}_P$ is the lower dimensional vector obtained by taking only the components of $\boldsymbol{x}$ of the positions of M, or equivalently, removing the components of $\boldsymbol{x}$ of the positions that are in $N - M$. For $M \subset N$ and $\boldsymbol{v} = (v_1, \ldots, v_n) \in S$, the *M*, ***v**-restricted bargaining problem* is the lower m-dimensional bargaining problem $(S_M(\boldsymbol{v}), \boldsymbol{u}_{0M}) = \{\boldsymbol{u}_M : \boldsymbol{u} = (u_1, \ldots, u_n) \in S \text{ and } u_k = v_k \text{ for all } k \notin S\}$.[41] Figure 25 illustrates a $\{1,2\}$, $(\frac{1}{2}, \frac{1}{2}, \frac{1}{2})$-restriction of the Figure 4 Wine Division Problem. In restricted problem $(S_{\{1,2\}}(\frac{1}{2}, \frac{1}{2}, \frac{1}{2}), \boldsymbol{0}_2)$, the German (Agent 1) and the Frenchman (Agent 2) engage in the two-dimensional problem that results from the original three-dimensional problem by holding the Spaniard's (Agent 3's) payoff fixed at $u_3 = \frac{1}{2}$.

For varying numbers of agents, there is an associated property analogous to Strong Monotonicity stated by the axiom

> *Population Monotonicity*: If $M \subset N$ and $\boldsymbol{f}(S, \boldsymbol{u}_0) = \boldsymbol{u}^*$, then $\boldsymbol{f}(S_M(\boldsymbol{u}_0), \boldsymbol{u}_{0M}) \geqq \boldsymbol{u}^*_{\mathrm{M}}$.

40 Years before Egalitarian and the Leximin solutions were first axiomatized, Luce and Raiffa showed with a different set of examples that Pareto Optimality and Strong Monotonicity can come into conflict. Luce and Raiffa considered a pair of essential but trivial bargaining problems embedded in a larger triangular-shaped bargaining problem (1957, 133–134).

41 In their treatments of axiomatic bargaining where the number of agents can vary, Thomson and Lensberg (1989) and Peters (1992) allow both $M \subset \mathbb{N}$ and $N \subset \mathbb{N}$ to vary across the natural numbers $\mathbb{N}$ with the sole restriction that $M \subset N$. This allows them to analyze bargaining problems having potentially infinitely many agents.

Table 2 Comparison of axioms related to changes in the feasible set

	Nash $N(S, u_0)$	**Kalai–Smorodinsky** $K(S, u_0)$	**Egalitarian** $E(S, u_0)$	**Weighted Nash** $N^{\alpha}(S, u_0)$	**Maximin Proportionate Gain** $G(S, u_0)$	**Leximin** $L(S, u_0)$
Weak Mutual Advantage	X	X	X	X	X	X
Strong Mutual Advantage	X	X	X	X	X	X
Weak Pareto Optimality	X	X	X	X	X	X
Pareto Optimality	X			X	X	X
Scale Invariance	X	X		X	X	
Symmetry	X	X	X		X	X
Contraction Consistency	X		X	X		
Restricted Monotonicity		X	X		X	
Strong Monotonicity			X			
Decomposability			X			

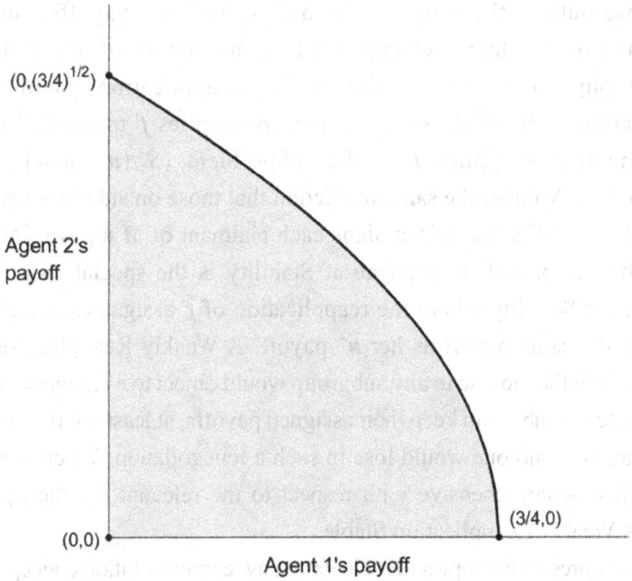

Figure 25 $\{1,2\},(\frac{1}{2},\frac{1}{2},\frac{1}{2})$-restriction of Hume's Wine Division Problem

Population Monotonicity says that if newcomers join in an original bargaining problem, then in the resulting new problem none of the original claimants gain merely on account of the newcomers' presence. Equivalently, Population Monotonicity says that if some in an original bargaining problem depart, then in the resulting new problem none of the remaining claimants lose merely on account of the departers' absence. When the feasible sets are all $\boldsymbol{u}_0$-comprehensive with respect to the relevant $\boldsymbol{u}_0$,[42] the Egalitarian and the Kalai–Smorodinsky solutions are Population Monotone. Two more axioms address how a solution might respond when reapplied across subgroups of the claimants:

Reapplication Stability: If $M \subset N$ and $\boldsymbol{f}(S, \boldsymbol{u}_0) = \boldsymbol{u}^*$, then $\boldsymbol{f}(S_M(\boldsymbol{u}^*), \boldsymbol{u}_{0M}) = \boldsymbol{u}^*_M$.

Weak Reapplication Stability: If $M \subset N$ and $\boldsymbol{f}(S, \boldsymbol{u}_0) = \boldsymbol{u}^*$, then $\boldsymbol{f}(S_M(\boldsymbol{u}^*), \boldsymbol{u}_{0M}) \geqq \boldsymbol{u}^*_{\mathrm{M}}$.

Reapplication Stability says that if one first applies the solution function $\boldsymbol{f}$ to the full problem $(S, \boldsymbol{u}_0)$ that assigns all of the claimants of N their payoffs $\boldsymbol{u}^* = (u^*_1, \ldots, u^*_n)$ and then applies $\boldsymbol{f}$ again to the subproblem $(S_M(\boldsymbol{u}^*), \boldsymbol{u}_{0M})$ of some subgroup $M \subset N$ while those outside M keep their $\boldsymbol{u}^*$ payoffs, then $\boldsymbol{f}$ again assigns the subgroup claimants their $\boldsymbol{u}^*$ payoffs. A Reapplication Stable solution is proof against renegotiation among those in some subgroup

[42] The nonagreement points can vary in that some can be M-restrictions of others.

when those outside the subgroup keep their assigned payoffs, since such a renegotiation would not change the assigned payoff of any claimant in the subgroup. The Nash solution is Reapplication Stable. Weak Reapplication Stability says that if one first applies f to the full problem $(S, \boldsymbol{u}_0)$ and then reapplies f to the subproblem $(S_M(\boldsymbol{u}^*), \boldsymbol{u}_{0M})$ of some subgroup $M \subset N$ under the same restriction that those outside M keep their $\boldsymbol{u}^*$ payoffs, the reapplication of f assigns each claimant of M a payoff at least as good as her $\boldsymbol{u}^*$ payoff. Reapplication Stability is the special case of Weak Reapplication Stability where the reapplication of f assigns each claimant of M exactly the same payoff as her $\boldsymbol{u}^*$ payoff. A Weakly Reapplication Stable solution is such that no one in any subgroup would object to a renegotiation when those outside the subgroup keep their assigned payoffs, at least not if all claimants are rational, since no one would lose in such a renegotiation. When the feasible sets are all $\boldsymbol{u}_0$-comprehensive with respect to the relevant $\boldsymbol{u}_0$, the Egalitarian solution is Weakly Reapplication Stable.

Here I will present examples that illustrate how certain solution concepts satisfy, or fail to satisfy, some of these subgroup-related stability axioms. Figure 26 depicts a three-agent bargaining problem $(S, \mathbf{0}_3)$ together with the reduced two-agent problem $(S_{\{1,2\}}(\mathbf{0}_3), \mathbf{0}_2)$ obtained by removing Agent 3.[43] In the reduced two-

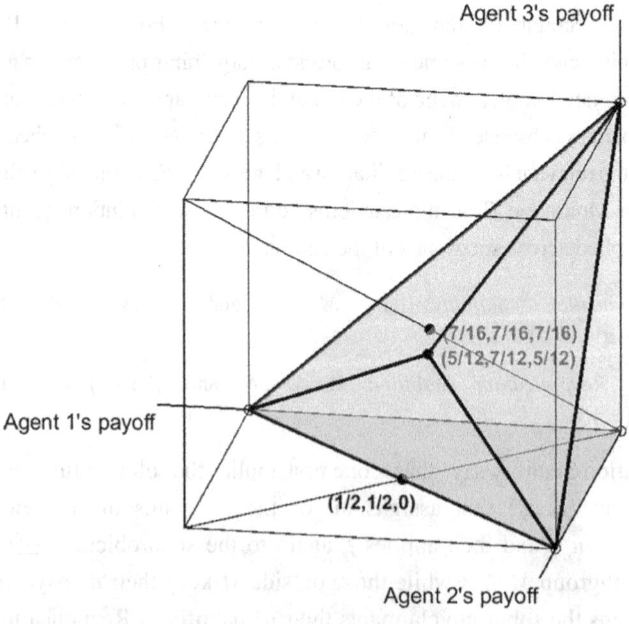

Figure 26 Population Monotonicity example

[43] This example is due to Thomson and Lensberg (1989, 41–42).

agent problem the Nash, Egalitarian, and Kalai–Smorodinsky solutions all coincide at $\boldsymbol{N}(S_{\{1,2\}}(\mathbf{0}_3), \mathbf{0}_2) = \boldsymbol{E}(S_{\{1,2\}}(\mathbf{0}_3), \mathbf{0}_2) = \boldsymbol{K}(S_{\{1,2\}}(\mathbf{0}_3), \mathbf{0}_2) = \left(\frac{1}{2}, \frac{1}{2}\right)$. In the three-agent bargaining problem $(S, \mathbf{0}_3)$, the Egalitarian and Kalai–Smorodinsky solutions coincide at $\boldsymbol{E}(S, \mathbf{0}_3) = \boldsymbol{K}(S, \mathbf{0}_3) = \left(\frac{7}{16}, \frac{7}{16}, \frac{7}{16}\right)$ and the Nash solution is $\boldsymbol{N}(S, \mathbf{0}_3) = \left(\frac{5}{12}, \frac{7}{12}, \frac{5}{12}\right)$. If Agent 3 joins Agent 1 and Agent 2, then when the Egalitarian and the Kalai–Smorodinsky solutions are reapplied in the new three-agent problem neither of the original agents gains, and in fact both lose in this case. This illustrates the Population Monotonicity of the Egalitarian and the Kalai–Smorodinsky solutions. But when the Nash solution is reapplied to the new three-agent problem, original Agent 2 gains. This shows the Nash solution is not Population Monotone. In a pair of actual bargaining problems with these structures, one might expect Agent 1 to object to the Nash solution on the grounds that in the three-agent problem they lose while Agent 2 gains merely because Agent 3 has joined them.

In the Figure 4 Wine Division Problem, the Nash, Egalitarian, and Kalai–Smorodinsky solutions all coincide at $\boldsymbol{N}(\boldsymbol{W}, 0_3) = \boldsymbol{E}(\boldsymbol{W}, 0_3) = \boldsymbol{K}(\boldsymbol{W}, 0_3) = \left(\frac{1}{2}, \frac{1}{2}, \frac{1}{2}\right)$. Figure 27 depicts this solution point in the three-agent problem. If the Spaniard's (Agent 3's) solution payoff of $\frac{1}{2}$ is held fixed, then in the reduced two-agent bargaining problem $\left(W_{\{1,2\}}\left(\frac{1}{2}, \frac{1}{2}, \frac{1}{2}\right), \mathbf{0}_2\right)$ the Nash and Egalitarian solutions coincide at $\boldsymbol{u}^*_{\{1,2\}} = \left(\frac{1}{2}, \frac{1}{2}\right)$. So if the German (Agent 1) and

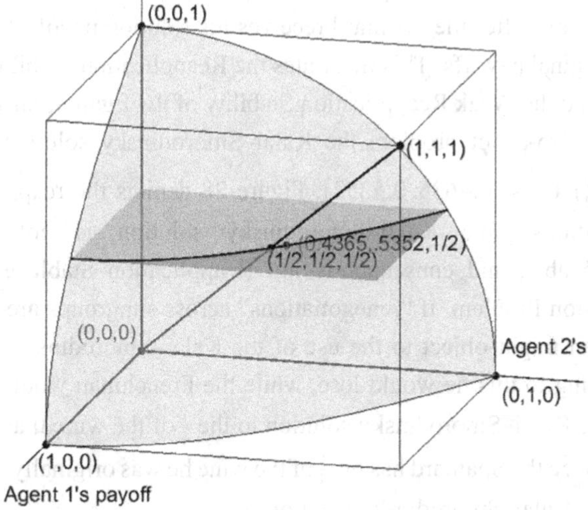

Figure 27 Nash/Egalitarian/Kalai–Smorodinsky solution of three-agent Wine Division Problem

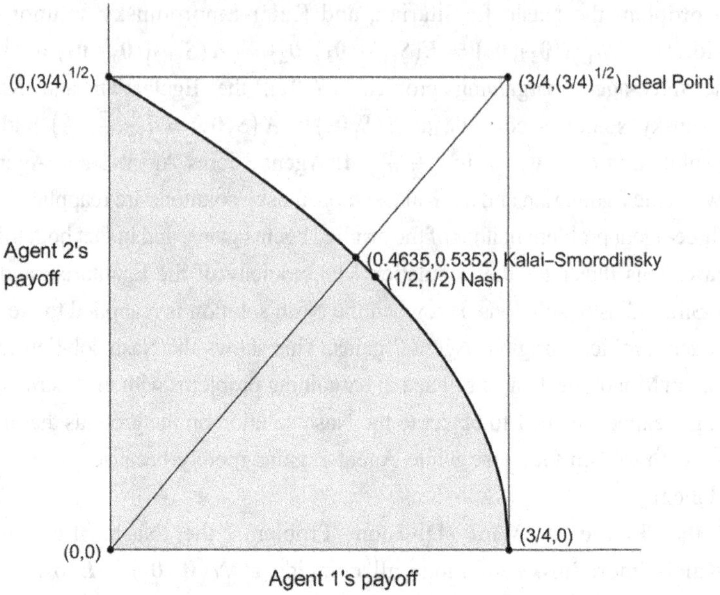

Figure 28 Reapplications of solutions to the $\{1,2\},(\frac{1}{2},\frac{1}{2},\frac{1}{2})$-restriction of Hume's Wine Division Problem

the Frenchman (Agent 2) "renegotiate" on the basis of the Nash or the Egalitarian solutions after the Spaniard receives his solution payoff, then they receive their original payoffs. This illustrates the Reapplication Stability of the Nash solution and the Weak Reapplication Stability of the Egalitarian solution. In the reduced two-agent problem the Kalai–Smorodinsky solution is now $\boldsymbol{K}(W_{\{1,2\}}(\frac{1}{2},\frac{1}{2},\frac{1}{2}),\mathbf{0}_2) \approx (0.4635, 0.5352)$. Figure 28 depicts the reapplications of these solutions. The Kalai–Smorodinsky solution is not Weakly Reapplication Stable, and consequently not Reapplication Stable either. In this Wine Division Problem, if "renegotiations" across subgroups are allowed then the German might object to the use of the Kalai–Smorodinsky solution because according to this he would lose, while the Frenchman would gain, if they reapply the Kalai–Smorodinsky solution to the $\frac{3}{4}$ of the wine that remains between them once the Spaniard has his $\frac{1}{4}$ of the wine he was originally assigned according to the Kalai–Smorodinsky solution.

Table 3 complements Table 2 by summarizing which of the subproblem related axioms discussed in this section the solution concepts discussed in this Element satisfy over the domain of $\boldsymbol{u}_0$-comprehensive sets.

Table 3 Comparison of axioms related to adding and removing agents

	Nash $N(S, u_0)$	**Kalai–Smorodinsky** $K(S, u_0)$	**Egalitarian** $E(S, u_0)$	**Weighted Nash** $N^{\alpha}(S, u_0)$	**Maximin Proportionate Gain** $G(S, u_0)$	**Leximin** $L(S, u_0)$
Population Monotonicity		X	X			
Reapplication Stability	X			X		X
Weak Reapplication Stability	X		X	X		X

5.4 Further Strategic Analyses

While axiomatic analysis takes the perspective of an arbiter, the above discussion illustrates how specific axioms that are motivated by the idea of making a bargaining problem proof against certain renegotiations might be consistent with the strategic perspective. As Nash's original analysis illustrates, strategic bargaining analyses try to model a bargaining process directly. Perhaps the best-known such models are *alternating offers* models, the first of which Ingolf Stähl and Ariel Rubinstein introduced in their pioneering studies (1972, 1982). In an alternating offers model agents take turns in different roles in a game that bears some resemblance to a Nash demand game but is in *extensive form* – that is, this game has an explicit sequential structure. At a given time period one agent is in the role of *Proposer* and the others are in *Recipient* roles. The Proposer moves first by making a *proposal* that specifies a division of the goods at stake. The Recipients move next after they have learned this proposal, and each Recipient then either *accepts* or *rejects* the proposal. If the Recipients all accept, then each receives the share specified. But if any Recipients reject, then they all receive nothing and at the next period the agents may repeat this extensive form game with a different agent taking their turn as Proposer. In Stähl's and Rubinstein's original alternating offers models, two agents take turns at successive periods as Proposer and each Agent *i* discounts the payoff they might eventually receive by her discount factor $\beta_i \in (0, 1)$. They show that if such a pair of agents have common knowledge of their rationality, the structure of their repeated game, and their discount factors, they will converge to a unique solution that is a *subgame perfect equilibrium* – that is, an equilibrium where agents follow only rational moves at every possible period. In Rubinstein's model the agents have a fixed and infinitely divisible good at stake. Rubinstein, Ken Binmore, and Asher Wolinsky (1986) show that for two variations of the Rubinstein model, as the length of the successive time periods approaches zero the solution the agents follow converges to the $\boldsymbol{\alpha}$-weighted Nash solution $(u_1 - u_{01})^{\alpha_1}(u_2 - u_{02})^{\alpha_2}$ where α_i is defined as an increasing function of β_i (Binmore, Rubinstein, and Wolinsky, 1986).[44] As discussed in Section 4, one can interpret $\alpha_i > \alpha_j$ as reflecting Agent *i* being more patient than Agent *j* in the bargaining process, so that Agent *i* has somewhat greater bargaining power than Agent *j* on this dimension. One might conclude that the Rubinstein model together with some of its variations vindicate this idea, at least for this special class of two-agent problems where the agents satisfy the strong stated common knowledge

44 Binmore was the first to recognize the equivalence of the Rubinstein strategic model solution and the $\boldsymbol{\alpha}$-weighted and ordinary Nash solutions. Binmore first presented this finding in a 1980 discussion paper later published in Binmore and Dasgupta (1987), 61–76.

assumptions. Some also conclude that the Rubinstein model lends support to the $\boldsymbol{\alpha}$-weighted Nash solution in general and to the ordinary Nash solution in particular if one supposes that $\alpha_1 = \alpha_2$ so that the strategic solution approaches a solution satisfying Symmetry.

Binmore, Rubinstein, and Wolinsky's results mark a tremendous step forward in completing the Nash program. However, alternating offers models have their limitations. Some extensions of the Rubinstein model applied to an n-agent bargaining problem where $n \geq 3$ and $\boldsymbol{\alpha} = (\alpha_1, \ldots, \alpha_n)$ can fail to converge to the $\boldsymbol{\alpha}$-weighted Nash solution. Indeed, Rubinstein and Martin Osborne show by example that a three-agent extension of the two-agent Rubinstein model can converge to any of the Pareto optimal payoff vectors (Osborne and Rubinstein 1990, 63–65).[45] So for a three-agent bargaining problem, this particular alternating offers model can fail to serve even to narrow the set of possible solutions in any interesting way other than their being Pareto optimal. And Eric van Damme, Reinhard Selten, and Eyal Winter (1990) show that even in the two-agent case, if the good at stake is a good like money that is divisible only up to finitely many increments, then two agents who otherwise satisfy the assumptions of the Rubinstein model might settle into practically any outcome better for both than the nonagreement point. To be sure, some other strategic models can produce more determinate results. Vijay Krishna and Roberto Serrano (1996) present another variation on the Rubinstein model where at each period if only a proper subset of the Recipients reject the current proposal, then the accepting Recipients are permitted to exit with their proposal payoffs so that only the current Proposer and the rejecters continue in the sequence of alternating offers. They show that this sequence of games has a unique subgame perfect equilibrium. They show further that if the agents all have the same discount factor β then this equilibrium converges to the ordinary Nash solution as $\beta \to 1$. Krishna and Serrano's alternating offers model clearly reflects the Reapplication Stability axiom and is another fine illustration of the interplay between axiomatic and strategic bargaining theory. Their model together with the three-agent model Rubinstein and Osborne explore also show how sensitive the resulting outcome of strategic analysis can be to the specific assumptions of the strategic model.

Other strategic models produce the Egalitarian, Leximin and Kalai–Smorodinsky solutions. I summarize two examples here. Walter Bossert and Guofu Tan (1995) propose a strategic model where for up to a finite number of time periods two agents engage in a demand game. The demand game

[45] Osborne and Rubinstein credit the example and the three-agent extension of the Rubinstein model to Avner Shaked.

resembles Nash's demand game in that Agent 1 and Agent 2 issue simultaneous respective claims $v_1(T)$ and $v_2(T)$ with respect to the nonagreement point $\boldsymbol{u}_0(T) = (u_{01}(T), u_{02}(T))$ at a time period T where $1 \leq T \leq M$, and each receives their claim if $v_1(T)$ and $v_2(T)$ are compatible, in which case their interaction ends. But in Bossert and Tan's model, if Agent 1 and Agent 2 issue incompatible claims at any period T before they exceed their limit of engagements M, they reengage in another demand game where if $v_1(T) - u_{01}(T) > v_2(T) - u_{02}(T)$ then the nonagreement point is now $\boldsymbol{u}_0(T+1) = (u_{01}(T), v_2(T))$, and similarly if $v_2(T) - u_{02}(T) > v_1(T) - u_{01}(T)$ then $\boldsymbol{u}_0(T+1) = (v_1(T), u_{02}(T))$. That is, the agent who issued the lower of the incompatible claims from the current nonagreement point is rewarded for being less demanding by having the non-agreement point moved in their favor. And if they exceed their engagement limit M, then they receive their respective payoffs of the original $\boldsymbol{u}_0(1)$. Bossert and Tan assume the two agents have common knowledge of their rationality and the structures of the demand games according to this process. They show that if $(S, \boldsymbol{u}_0(1))$ is strictly $\boldsymbol{u}_0$-comprehensive then the agents follow the Egalitarian solution $\boldsymbol{E}(S, \boldsymbol{u}_0(1))$ and that if $(S, \boldsymbol{u}_0(1))$ is only $\boldsymbol{u}_0$-comprehensive then the agents follow the Leximin solution $\boldsymbol{L}(S, \boldsymbol{u}_0(1))$.

Herve Moulin (1984) presents a strategic model where the agents bid for the right to propose an agreement payoff vector for them all to follow. Moulin defines the agents' bargaining problem in terms of a basis game where a given claim profile $\boldsymbol{x} = (x_1, \ldots, x_n)$ defines a given payoff vector $\boldsymbol{u}(\boldsymbol{x})$. Moulin also assumes the feasible set is $\boldsymbol{u}_0$-comprehensive and that the nonagreement point is $\boldsymbol{u}_0 = \boldsymbol{u}(\boldsymbol{0})$ where each agent gets nothing.[46] Moulin normalizes the payoff scales so that for each Agent i, their nonagreement point payoff at an outcome where they get nothing is $u_{0i} = u_i(\boldsymbol{0}) = 0$ and their ideal payoff at the outcome where they get everything is $\overline{u}_i = 1$.[47] In Moulin's game, at an initial stage each Agent i independently submits a bid $p_i \in (0, 1)$, and Agent i_1 who has submitted the highest bid p_1 becomes the *Proposer*.[48] Agent i_1 completes the first stage by proposing a claim profile $\boldsymbol{x}^*$ that would yield $\boldsymbol{u}(\boldsymbol{x}^*)$ as their solution. Then in turn each of the other agents may adopt the role of *Receiver*. If the game reaches the kth stage for $k > 1$, then Agent i_k takes her turn as Receiver, and then either *accepts* or *rejects* the proposed $\boldsymbol{x}^*$. If Agent i_k accepts, the game continues on to the $k+1$ st stage and the next Agent i_{k+1} in line takes his turn as Receiver. If

[46] Moulin describes this basis game construction in Moulin 1984, 33–35. This construction generates a $\boldsymbol{u}_0$-comprehensive set although Moulin does not state this explicitly.

[47] The claim profile where Agent i receives all of the good at stake and the others none is the unit vector $\boldsymbol{e}_i \in \mathbb{R}^n$ where the ith coordinate is 1 and all the other coordinates are 0, so in Moulin's bargaining problem $\overline{u}_i = u_i(\boldsymbol{e}_i) = 1$.

[48] Moulin notes that in case more than one agent submits the highest bid, the Proposer agent can be selected according to some random device (1984, 37).

Agent i_k rejects, then with probability p_1 the agents follow the outcome where Agent i_k gets everything and the others nothing, and with probability $1 - p_i$ they follow the nonagreement point. So if Agent i_k rejects, then her expected payoff is $\overline{u}_{i_k}p_1 + (1 - p_1)u_{0i_k} = p_1$. Hence by increasing the bid p_i each Agent i increases their chance of winning the Proposer position, but at the same time increases the attractiveness for each of the others of rejecting their own proposed solution in hopes of achieving their ideal outcome and leaving the others with nothing. Moulin shows that if the agents have common knowledge of this procedure, then the unique subgame perfect equilibrium is the Kalai–Smorodinsky solution.

5.5 Lessons from the Rational Choice Program

In the end, the rational choice approach may be less about choosing one solution concept for all bargaining problems and more about clarifying the properties one regards as essential for solving a given bargaining problem or set of bargaining problems. As we have seen in this section, the results of both the strategic and the axiomatic approaches are sensitive to the choices of models and axioms. This phenomenon is general. A fundamental lesson of the rational choice program is that in the analysis of bargaining problems one must decide which formal limitations to accept. More specifically, one must decide which, if any, limitations one must set upon feasible sets, and one must decide which formal desiderata one will require and which one will sacrifice. The Figure 21 example shows that if one regards Weak Mutual Advantage, Pareto Optimality, Scale Invariance, and Symmetry as essential formal properties, then one apparently must choose between Restricted Monotonicity and Contraction Consistency, an application of Sen's Property α. The Figure 24 example shows that if one does not require feasible sets to be strictly $\boldsymbol{u}_0$-comprehensive, then if one insists upon Weak Mutual Advantage, Symmetry, and Contraction Consistency one apparently must still choose between Pareto Optimality and Strict Monotonicity, even if one is already prepared to give up Scale Invariance. And so on. At perhaps an opposite extreme, the Egalitarian solution satisfies almost all of the stability properties discussed in this section if one limits the domain to the strictly $\boldsymbol{u}_0$-comprehensive sets. For this domain a solution $\boldsymbol{f}(S, \boldsymbol{u}_0)$ satisfies Pareto Optimality, Symmetry, and Strict Monotonicity exactly when $\boldsymbol{f}(S, \boldsymbol{u}_0) = \boldsymbol{E}(S, \boldsymbol{u}_0)$.[49] On this domain the Egalitarian solution is also Decomposable, Contraction Consistent, Population Monotone, and Reapplication Stable. The Egalitarian solution "pays" for these excellent

[49] Peters (1992, 77) and Roemer (1996, 74–75) give elegant proofs of this result, which is an extension of a closely related theorem Kalai originally proved (1977, 1626–1627).

stability properties over this domain both by the very fact that the domain is limited to the strictly $\boldsymbol{u}_0$-comprehensive sets and by its implicit use of interpersonally comparable utilities. One gains so many other stability properties over this domain by giving up Scale Invariance. And if one expands the domain only "moderately" to include all $\boldsymbol{u}_0$-comprehensive feasible sets, then the Egalitarian solution is only Weakly Pareto Optimal and only Weakly Reapplication Stable.

Examples such as those discussed above can help illuminate what is at stake in adopting or relaxing certain formal properties summarized by given axioms, and perhaps also show how certain axioms appear, or do not appear, to be compelling in certain contexts. To illustrate, the seemingly simple Figure 22 example has sparked an interesting discussion over the relative merits of certain axioms. Some argue that this should throw our intuitions regarding Strict Monotonicity and even Restricted Monotonicity into doubt since these properties can apparently clash with Property α, which when applied to a bargaining problem requires a solution to be in some sense a maximal element of the feasible set. The Nash solution identifies such a maximal element since this solution is defined as the vector $\boldsymbol{u} \in S$ that maximizes the Nash product with respect to $\boldsymbol{u}_0$. Others question adherence to Property α in this context. They point out that in this pair of bargaining problems, the expanded $(S_1, \boldsymbol{u}_0)$ introduces new payoff vectors that increase only Agent 2's possible utility levels in the larger S_1 for this scaling. And the ideal point of the expanded $(S_1, \boldsymbol{u}_0)$ changes from that of the smaller $(S_0, \boldsymbol{u}_0)$. Why should Agent 2 not gain from an expansion apparently in Agent 2's favor only? Those who defend Restricted Monotonicity over Contraction Consistency in this context can also argue that Strong Monotonicity is not so compelling in this context, where the expansion of S_0 does not introduce payoff vectors that even weakly dominate any of those in S_0.

Some even challenge Nash's idea that a solution concept must always select a unique payoff vector. Many solution concepts, including most of the concepts discussed in this Element, require Symmetry. Without Symmetry, the resulting generalizations of the Egalitarian and Kalai–Smorodinsky solutions no longer select unique payoff vectors, and the asymmetric $\boldsymbol{\alpha}$-weighted Nash Solution is well defined only when $\boldsymbol{\alpha}$ is specified. John Harsanyi points out that the Symmetry axiom reflects Nash's assumption that one sort of bargaining solution has a determinate solution, namely a symmetric bargaining problem (1977, 144). Nash then shows that if one adopts all of his axioms, including Symmetry, then all the essential bargaining problems have a determinate solution. But Schelling (1960, 278–284) and more recently John Thrasher (2014) question Nash's assumption. As part of a more general critique of the rational choice approach to analyzing bargaining problems, Schelling argues that while

Symmetry is compatible with rationality, to incorporate Symmetry into the definition of rationality begs the question. Harsanyi himself acknowledges that Nash's Symmetry axiom is not part of classical economics or von Neumann and Morgenstern's theory of Bayesian rationality (1977, 144). Thrasher argues that Symmetry is actually a normative constraint that begs the question in favor of certain solution concepts, including in particular those concepts that emulate Nash's idea that a solution concept defines a function over the domain of bargaining problems. Schelling and Thrasher maintain that in at least some contexts, one should be willing to relax the assumption in so much of bargaining theory that every bargaining problem has a unique solution rational agents will follow. Relaxing this assumption opens the door to the possibility of *bargaining conventions*, the topic of the following section.

6 Bargaining Conventions

Hume argues that three claimants can resolve their dispute over a quantity of wine by following a convention that defines a specific partition of the wine. Hume's Wine Division example illustrates a general conventionalist approach to analyzing bargaining problems that complements the rational choice approach in interesting ways. As we have seen in Section 5, the motivating idea of the rational choice approach to analyzing bargaining problems is that rational and sufficiently knowledgeable agents can identify a unique outcome as the solution. The axiomatic side of this approach presupposes the arbiter is rational and knows the full payoff structure of the bargaining problem so that they can assign claimants their individual solution payoffs. The strategic side of this approach typically presupposes the claimants have common knowledge of rationality and feasible set payoffs. One might doubt that either of these sets of epistemic assumptions clearly reflect what the agents who are arbiters or claimants in actual bargaining situations know regarding their situation. Indeed, as R. Duncan Luce and Howard Raiffa (1957, 134) and Thomas Schelling (1960, 23) recognize, in many actual bargaining situations the agents in the positions of claimants might be tempted to try to conceal their rationality, their true payoffs, or their willingness to make certain concessions from one another or from their arbiter. The motivating idea of the conventionalist approach to analyzing bargaining problems is that the agents who engage in a bargaining problem can settle into one of many distinct available bargaining conventions by making the right inferences from their common background, including what they learn from repeated interaction. The conventionalist approach does not presuppose that a given bargaining problem must have a unique solution, and the epistemic assumptions of this approach are quite different from those of the rational choice approach.

One can define a *basic convention* as a strict correlated equilibrium of a game in which a community of agents engage that has a plurality of such equilibria together with these agents having common knowledge of these facts. A convention is *incumbent* if the agents of this community also have common knowledge that they expect each other to follow the equilibrium of *this* convention, rather than any of the alternative convention equilibria (Vanderschraaf 2019, 81–83). This game-theoretic definition of conventions generalizes Lewis' own definition and is similar to Sugden's generalization of Lewis' definition (Lewis 1969, 78–79; Sugden (1986) 2004, 33–35). The underlying idea of such game-theoretic analyses is that a convention is some pattern of social practice that the members of a community recognize as mutually beneficial and will follow as *their* convention on condition they expect each other to follow *this* practice. As noted in Section 2, this analysis is quite similar to Hume's analysis of convention. In Hume's Wine Division Problem, each partition that allots each claimant a positive quantity of the available wine characterizes a convention. But Hume thinks that an arbiter would implement a particular convention, one that assigns to each claimant the bottle of wine produced in his home country: "And this from a principle, which in some measure, is the source of those laws of nature, that ascribe property to occupation, prescription and accession" ((1740) 2000, 3.2.3: 10 n. 5). Hume's recommended convention is an *accession* convention, in that each part of the corresponding equilibrium is in a sense "near" the given claimant. Hume supposes one would naturally link together the geographical origins of the wines and the claimants, given that these are the only stated properties that the wine and the person obviously have in common.

Some of the early pioneers of game theory, including Luce and Raiffa (1957, 105) and Nash himself ((1951b) 1996, 32–33), argued that agents who engage in a game repeatedly could gradually converge to equilibria as the result of what they learn through these repeated engagements.[50] In *The Strategy of Conflict*, Schelling argues that agents who engage in coordination problems, including bargaining problems, can coordinate their actions on equilibrium outcomes far more often than random chance would predict, by inferring how to act in accord with clues from the context of their interaction that they all recognize.

> Finding the key, or rather finding *a* key – any key that is mutually recognized as the key becomes *the* key – may depend on imagination more than on logic; it may depend on analogy, precedent, accidental arrangement, symmetry,

[50] This discussion is the last part of Nash's doctoral thesis. Nash (1951a) is a revised version of his doctoral thesis that omits this last discussion.

> aesthetic or geometric configuration, casuistic reasoning, and who the parties are and what they know about each other. (1960, 57)[51]

Remarkably, these inductive learning and mutually recognized key approaches to equilibrium selection in games are both foreshadowed in Hume's analysis of convention.[52] Hume's solution of the Wine Division Problem is an example of what Schelling calls a *focal point* and what Lewis calls a *salient* equilibrium (1960, 57–58; 1969, 35–36). Schelling's and Lewis' works launched a contemporary research program that explores how equilibria in games might emerge because they somehow "stand out." Focal point explanations of efficient allocations in bargaining problems are illustrated in many examples in ordinary life and in the laboratory. In unstructured bargaining experiments, where the details of the bargaining process are left to the subjects themselves, subjects try to follow certain divisions they regard as focal. For example, in a classic experimental study designed by Alvin Roth and Keith Murnighan, pairs of subjects bargained over 100 lottery tickets that determined chances of one of them winning a monetary prize, where one pair member could win $20 and the other could win $5. They varied treatments by having each member of a pair know the value of only their own prize or the values of both prizes, and by making knowledge of the two prizes common knowledge. When neither pair member knew the value of their partner's prize, pairs tended to settle upon equal division of the tickets, which gave them equal chances of winning. When the values of both prizes were common knowledge, pairs tended to split between an equal division of the tickets and a division that produced equal expected monetary winnings (Roth and Murnighan 1982). Roth argues that the subjects in this study tended to follow these outcomes because for them the divisions of equal chances of winning and equal expected monetary winnings were focal points (Roth 1985).[53]

What makes a focal point focal? Schelling and Hume before him frankly acknowledge that the agents engaged in a coordination problem tend to employ different sets of contextual clues in different situations, and that such agents might in the end settle upon some outcome for reasons that seem frivolous to an outsider and perhaps even to themselves (Schelling 1960, 54–58; Hume (1740)

---

51 Schelling develops his views on coordination and focal points primarily in *The Strategy of Conflict* (1960), 81–203.

52 Hume gives his best discussions of these two approaches in *Treatise* 3.1.1–3 and *Enquiry* 3.2. Robert Sugden together with several of his colleagues have shown that Hume's insights into how individuals might settle into conventions according to contextual clues not only foreshadows this contemporary research but has much to teach the researchers (Mehta, Starmer, and Sugden 1992, 1994a, 1994b; Bardsley et al. 2010).

53 Roth (1995) and Camerer (2003, 151–198) give fine overviews of some of the best-known results of bargaining experiments and in particular how focal point effects can explain them.

2000, 3.2.3: 4 n. 1; Hume (1751) 1998, 3.2: 35–37). Interestingly, Schelling speculates that the mathematical properties of an axiomatic solution of a bargaining problem might capture the attention of the bargainers, so that this solution is focal for them, only if they perceive each other to be mathematicians, and possibly only if they also are game theorists (1960, 113–114). The bargaining problem as Nash formulates it might seem especially open-ended with respect to focal points given the huge number of strategy profiles, and accompanying payoff vectors, defined by various claim combinations available to the agents in any nontrivial bargaining problem. Experimental studies such as Roth and Murnighan's study illustrate how a bargaining problem could have multiple focal equilibria. In the Wine Division example, Hume regards his accession partition as salient, and claims the arbiter expresses his impartiality by implementing this partition. But Hume's arbiter apparently could also demonstrate his impartiality by implementing another partition, namely the partition that assigns each of the three claimants one third of the contents of each bottle. This *equal shares* solution evidently has some salience of its own, as this is the only partition of the three vintages that gives each claimant equivalent bundles of the goods at stake. Moreover, one might suspect that the claimants would regard the equal shares solution more salient than the accession solution if the bottles of Rhenish, Burgundy, and Port differed significantly in size, so that at least one bottle contained far more wine than one of the other bottles. Indeed, the actions of an outsider may be an important source of salience. I suspect that Hume introduces an arbiter into the Wine Division Problem because the arbiter can interrupt the claimants' quarrel and help focus their reciprocal expectations. Hume's arbiter distributes the wine himself, and Hume evidently thinks the claimants will find the accession division mutually acceptable so that this division becomes their convention. If this is so, then the arbiter might well have achieved an equivalent result just by recommending the accession solution and then allowing the claimants to take their recommended bottles. The claimants are of course capable of rejecting any partition the arbiter implements or recommends and continuing their quarrel. Such a suboptimal outcome might well result if the arbiter implements or proposes some partition that one or more of the claimants regards as unacceptable. But as Hume's example illustrates, the arbiter's activities can raise the prominence of some partition the claimants do find mutually acceptable so that this partition characterizes their bargaining convention.

In principle, any outcome of a bargaining problem could be salient on account of any property associated with this outcome only. Again in principle, all outcomes could be equally conspicuous, so that *none* are salient, because the bargainers have common knowledge they can tell all these outcomes apart and consider no other common background knowledge. But as already noted, empirical evidence confirms that in many bargaining problems the agents involved do use focal points to coordinate on efficient divisions. Some outcomes of some bargaining problems do "stand out." What could explain the origins of the focal point effect? As hinted above, one such explainer may be what agents learn from repeated interactions. Any focal point effect in a coordination situation depends crucially upon what the agents trying to concert their actions know about each other. Trial and error learning can serve as one source of this knowledge. Hume understands this in an informal manner, when he acknowledges the possibility that a rule of stability of possession "arises gradually, and acquires force by a slow progression, and by our repeated experience of the inconveniences of transgressing it" ((1740) 2000, 3.2.2: 11). In bargaining problems, agents can learn to follow certain types of equilibria more often than others even when they initially have no contextual information other than their own individual payoffs, and the results of the learning process can lead them to regard certain solutions as focal.

Here I will give examples showing how a focal point effect could emerge from learning. More specifically I will apply a simple model of learning, the *replicator dynamic*, to some specific bargaining problems. The replicator dynamic is perhaps the best-known dynamical model of cultural evolution among game theorists. This dynamic models a process of the evolution over time of the distribution of pure strategies represented in a population. According to the replicator dynamic, the proportion x_i of a population that currently follows a pure strategy $\boldsymbol{s}_i$ grows, or declines, according to how the current expected payoff of $\boldsymbol{s}_i$ compares with the overall average payoff of the population according to the current pure strategy distribution or *population state* $\boldsymbol{x} = (x_1, \ldots, x_n)$. A special case of a population state is $\boldsymbol{e}_i = (0, \ldots, 0, 1, 0, \ldots, 0)$ – that is, all members of the population follow $\boldsymbol{s}_i$. Over time, the proportions of the pure strategies that have higher expected payoffs than overall average payoffs increase, while the proportions of the pure strategies that have lower expected payoffs than overall average payoffs decrease.[54] One can interpret the replicator dynamic as a rule for learning by imitation, where population members have a tendency to adopt the strategies

[54] The laws of motion for the one-population replicator dynamic for a symmetric two-player game with $M+1$ pure strategies $s_0, \ldots, s_M$ are defined by the $M+1$ differential equations with respect to time t of the form

$$\frac{dx_i}{dt} = x_i \cdot (u(\boldsymbol{e}_i, \boldsymbol{x}) - u(\boldsymbol{x}, \boldsymbol{x}))$$

where $\boldsymbol{x} = (x_1, \ldots, x_n)$ is current population state, $u(\boldsymbol{e}_i, \boldsymbol{x})$ is the payoff a pure strategy $\boldsymbol{s}_i$-follower gains given state $\boldsymbol{x}$, and $u(\boldsymbol{x}, \boldsymbol{x})$ is the average population payoff at state $\boldsymbol{x}$. The laws of motion for the two-population replicator dynamic discussed below are defined similarly.

that are more successful in terms of their average payoffs. The epistemic assumptions of learning models like the replicator dynamic do not include common knowledge of rationality, payoffs, and other agent-specific parameters. Given the imitation learning interpretation, agents who update according to the replicator dynamic need know only their own payoffs in the game and the current frequency distribution of pure strategies present in the population in order to update their strategies. A population state $\boldsymbol{x}^*$ can be a *rest point* of the replicator dynamic, and $\boldsymbol{x}^*$ can be an *attracting point* in the sense that starting from any of some set of population states sufficiently near $\boldsymbol{x}^*$, this dynamic will converge to $\boldsymbol{x}^*$. The population states that define Nash equilibria are a proper subset of the replicator dynamic rest points.[55] Population states defining strict Nash equilibria are replicator dynamic attracting points, though not necessarily the only attracting points. The *basin of attraction* $\Delta_{\boldsymbol{x}^*}$ of an attracting point $\boldsymbol{x}^*$ of the replicator dynamic is the set of population states where if the dynamic starts from some $\boldsymbol{x} \in \Delta_{\boldsymbol{x}^*}$ then it converges to $\boldsymbol{x}^*$. The *size* $\mu(\Delta_{\boldsymbol{x}^*})$ of this basin is the fraction of the simplex Δ of all possible population states that $\Delta_{\boldsymbol{x}^*}$ defines.

For the bargaining problems discussed below, I will apply versions of the replicator dynamic to a variety of discrete demand games, each of which approximates with *claim precision* $\delta = \frac{1}{M}$ the corresponding demand game where each agent can claim any fraction of the good. In the approximating discrete demand game the good at state is effectively divided into M pieces of equal size δ, and each agent may claim some fraction $m_i\delta$ of the good where $m_i \in \{0, 1, \ldots, M\}$. If those in the roles of Agent 1 and Agent 2 issue respective compatible claims m_1 and m_2, then they achieve the payoff vector $(u_1(m_1\delta, m_2\delta), u_2(m_1\delta, m_2\delta))$, and otherwise they receive their nonagreement point payoffs. Each claim pair $(m\delta, (M-m)\delta)$ where $m \in \{1, \ldots, M-1\}$ is a strict equilibrium and hence characterizes both a convention of the discrete demand game and a replicator dynamic attracting point. But different convention equilibria need not all have the same attracting power. Moreover, convention equilibria are not necessarily the only attracting points of these dynamics.

Figure 29 summarizes a two-agent game that approximates the demand game of the Figure 1 Chocolate Cake Problem with $\delta = \frac{1}{2}$. In this game, the cake is divided into two equal-sized pieces, and each agent can claim nothing (s_0), one piece (s_1), or both pieces (s_2). This game has three basic conventions in pure

[55] In fact, the distribution of every pure strategy profile is a rest point of the replicator dynamic. However, only the distribution of a Nash equilibrium can be *Lyapunov stable* – that is, from no small change in this distribution does the replicator dynamic move away from this point. A replicator dynamic attracting point, sometimes also called an *asymptotically stable* point, is a Lyapunov stable point with the additional property that for any sufficiently small change in the distribution the replicator dynamic returns to this point.

		Agent 2		
		s_0	s_1	s_2
Agent 1	s_0	$(0,0)$	$(0,\frac{1}{2})$	$(0,1)$
	s_1	$(\frac{1}{2},0)$	$(\frac{1}{2},\frac{1}{2})$	$(-1,-1)$
	s_2	$(1,0)$	$(-1,-1)$	$(-1,-1)$

s_0 = claim nothing, s_1 = claim one piece, s_2 = claim both pieces

Figure 29 Two-piece Chocolate Cake Game

strategies, namely (s_0,s_2), (s_1,s_1), and (s_2,s_0) with respective payoff vectors $(0,1)$, $(\frac{1}{2},\frac{1}{2})$, and $(1,0)$. Brian Skyrms applies a *one-population* version of the replicator dynamic to this game, according to which all the members of a single population have the same pure strategies and the same payoff matrix in each of their pairwise encounters ((1996) 2014, 14–15).[56] Figure 30 depicts a *phase diagram* of the single population replicator dynamic applied to this game. Each *orbit* of this phase diagram starts from an initial population state chosen at random from within the triangular simplex Δ of all possible population states over $S=\{s_0,s_1,s_2\}$. In this division game the population state $\boldsymbol{e}_1=(0,1,0)$ is a replicator dynamic attractor, with $\mu(\Delta_{\boldsymbol{e}_1})\approx 0.718$. At $\boldsymbol{e}_1$ all population members follow the convention (s_1,s_1) of equal division and achieve a payoff of $\frac{1}{2}$ in their encounters. However, Skyrms shows that this game has another replicator dynamic attractor. This attractor is the *polymorphic* population state $\boldsymbol{x}'=(\frac{1}{2},0,\frac{1}{2})=\frac{1}{2}\boldsymbol{e}_0+\frac{1}{2}\boldsymbol{e}_2$ – that is, half the population follow s_0 and half the population follow s_2, and $\mu(\Delta_{\boldsymbol{x}'})\approx 0.282$. $\boldsymbol{x}'$ characterizes the *mixed strategy* $\sigma_{\frac{1}{2}}=\frac{1}{2}s_0\oplus\frac{1}{2}s_2$ where a given agent in the Figure 29 game follows s_0 and s_2 with probability $\frac{1}{2}$ each, independently of the other agent's strategy. And at $\boldsymbol{x}'$ population members achieve an average payoff of only 0 in their encounters. Skyrms and Sugden before him show that s_1 and $\sigma_{\frac{1}{2}}$ are both *evolutionarily stable strategies* (Skyrms (1996) 2014, 10–11; Sugden (1986) 2004, 71). A strategy $\boldsymbol{s}^*$ is evolutionarily stable when if $\boldsymbol{s}^*$ is *incumbent* so that all in the population follow $\boldsymbol{s}^*$, then this incumbent $\boldsymbol{s}^*$ can repel a limited-scale invasion of any other strategy (Maynard Smith 1982, 10, 14). So while the bargaining convention (s_1,s_1) of fair division defined by the strategy s_1 can evolve via this

[56] The Figure 29 game is an equivalent rescaling of Skyrms' division game.

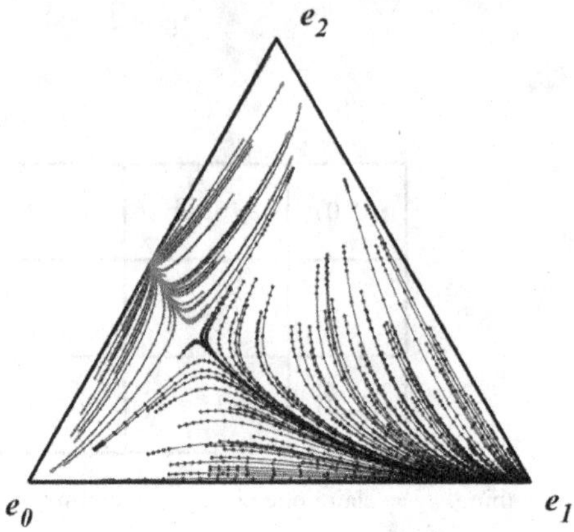

Figure 30 Phase diagram of replicator dynamic applied to Figure 29 game

sort of learning, so can a state equivalent to that where all follow $\sigma_{\frac{1}{2}}$ that is suboptimal and also fails to characterize a convention. And since $\sigma_{\frac{1}{2}}$ and s_1 are both evolutionarily stable, it is just as possible for the members of a population engaged in this division problem to remain "stuck" in a suboptimal polymorphism as it is for them to remain "stuck" in a bargaining convention.

But Skyrms considers a variation of the replicator dynamic that relaxes the assumption of the ordinary replicator dynamic that population members are matched together completely at random. In this *correlated replicator dynamic*, members of a population following particular strategies are somewhat more likely to meet counterparts following the same strategies. The level of correlation in encounters is governed by a parameter $\alpha \in [0, 1]$, where at $\alpha = 0$ encounters are completely random as in the ordinary replicator dynamic and at $\alpha = 1$ correlation is perfect, so that agents who follow a given strategy are always matched with partners who follow the same strategy. One way to interpret a level $\alpha > 0$ is to say that members of subgroups like families may tend to act alike, and also may tend to interact with each other more often than with those outside their own subgroups. Skyrms shows that a relatively modest level of correlation dramatically increases the prospects for the evolution of fair division in the Figure 29 game. Figure 31 depicts a phase diagram of the correlated replicator dynamic with $\boldsymbol{\alpha} = 0.20$ applied to the Figure 29 game.[57] For this level of correlation, all of the orbits now converge to the state $\boldsymbol{e}_1$ that characterizes s_1. The convention of equal division is now a

[57] In the ordinary replicator dynamic, x_i is the probability of meeting a pure strategy s_i-follower at time t as well as the proportion of the population that follow s_i at time t. In correlated replicator dynamics, $\alpha \in [0, 1]$ augments the probability at t of an s_i-follower meeting another s_i-follower, $x_{(i|i)}$, by the equation

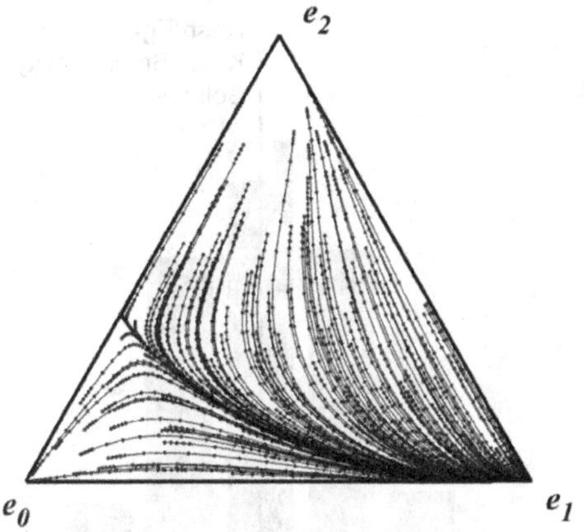

Figure 31 Phase diagram of correlated replicator dynamic with $\alpha = 0.20$ applied to Figure 29 game

global attractor – that is, from every initial state $\boldsymbol{x} \in \Delta$ the 0.20-correlated replicator dynamic converges to $\boldsymbol{e}_1$. In a subsequent study, Skyrms and Jason McKenzie Alexander show that agents who engage repeatedly in bargaining problems with their neighbors in a *network structure* and who learn according to an *imitation dynamic* where they adopt the strategies of their most successful neighbors tend to settle into equal division conventions (Alexander and Skyrms 1999). Studies such as these suggest a more general moral: Learning where there is some *sorting* in encounters may facilitate the emergence of certain conventions like an equal division bargaining convention.

I now turn to discrete demand games where each agent has a distinct role with associated payoffs. Such demand games can have asymmetric payoff structures. For these games I will apply a *two-population* version of the replicator dynamic, where representatives from two distinct populations engage in this game in the roles of Agent 1 and Agent 2. Figure 32 summarizes a more fine-grained approximating game of the Figure 1 Chocolate Cake Problem where $M = 10$, so that $\delta = \frac{1}{10}$, and where -1 is the nonagreement payoff. For this dynamic, the limits of the orbits starting from initial strategy distribution points chosen at random are distributed across the states of the set of convention equilibria. But

$$x_{(i|i)} = x_i + \alpha \cdot (1 - x_i)$$

and for pure strategy $\boldsymbol{s}_j \neq \boldsymbol{s}_i$ lowers the probability that an $\boldsymbol{s}_i$-follower meets an $\boldsymbol{s}_j$-follower, $x_{(i|j)}$, by the equation

$$x_{(i|j)} = x_i - \alpha x_j.$$

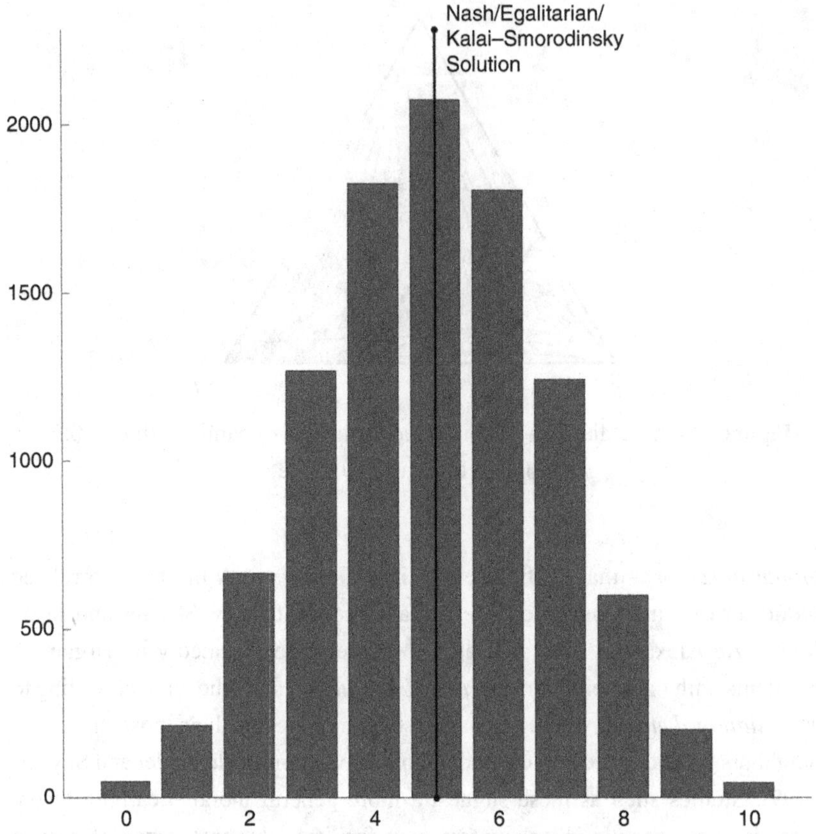

Figure 32 Relative sizes of basins of attraction of convention equilibria of the ten-piece Chocolate Cake Game

the state of the convention equilibrium (s_5, s_5) where each side claims half has the largest basin of attraction, and the distribution of orbit limits is centered around this convention equilibrium. This indicates that inductive learners who engage repeatedly in a structurally symmetric bargaining problem might settle into a variety of division conventions, *and* that they will have some tendency to converge to the equal division convention. This dovetails with everyday experience, where parties in a completely symmetric division problem might regard equal division as salient and follow the equal division equilibrium most of the time, but perhaps not always.

Given the symmetric payoff structure of the Figure 1 Chocolate Cake bargaining problem, the Nash, Kalai–Smorodinsky, and Egalitarian solutions all coincide at the $\left(\frac{1}{2},\frac{1}{2}\right)$ point of equal division. Do inductive learners tend to converge to equilibria at or near any of these axiomatic solutions in asymmetric bargaining problems? Figure 33 summarizes the relative sizes of the basins of attraction of the two-population replicator dynamic applied to a demand game that approximates the demand game of the Braithwaite Problem with claim precision $\delta = \frac{1}{10}$. In the Braithwaite Problem the Nash, Kalai–Smorodinsky, and Egalitarian solutions are mutually distinct. For the two-population replicator dynamic, the convention equilibrium of this game that most nearly approximates the Egalitarian solution of the Braithwaite Problem has the largest basin of attraction, followed by the convention equilibrium of the Kalai–Smorodinsky solution and then the convention equilibrium of the Nash solution. For this approximating game of the Braithwaite Problem, the Egalitarian solution has a certain attracting power for this form of inductive learning. As a more general test, I applied the two-population replicator dynamic to 1,000 approximating

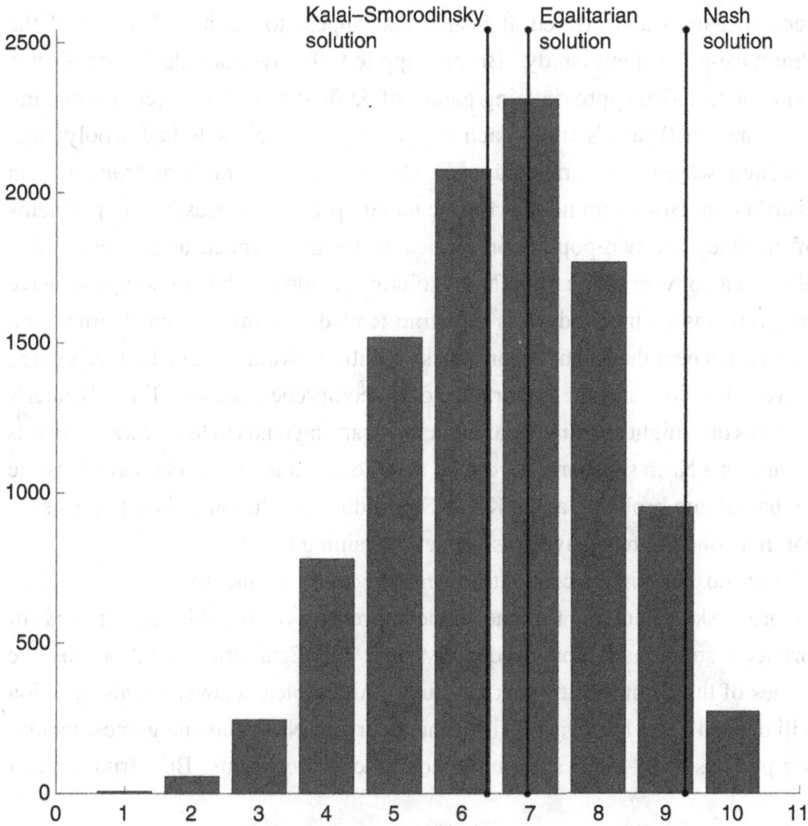

Figure 33 Relative sizes of basins of attraction of convention equilibria of the ten-piece approximating Braithwaite demand game

games with $\delta = \frac{1}{50}$, each constructed from one of 1,000 distinct conflictual coordination basis games that characterize a bargaining problem with the same ordinal payoff structure as that of the Braithwaite Problem. Details of the construction of these conflictual coordination games and the computer simulations of this *Braithwaite Ranking* simulation study are given in the Appendix. For these conflictual coordination games, the convention equilibrium with the largest basin of attraction was within δ of the Egalitarian solution 39.7 percent of the time, within δ of the Nash solution only 15.9 percent of the time, and within δ of the Kalai–Smorodinsky solution only 14.1 percent of the time.[58] For these sorts of bargaining problems, inductive learning may lead to the Egalitarian solution being somewhat focal.

Substantial work remains to be done exploring the conditions under which conventions that approximate well-known axiomatic solutions emerge, or fail to emerge, most often as the result of inductive learning. In one fine recent study Justin Bruner performed another set of computer experiments on a large set of demand games and arrived at results that appear to clash with those of the Braithwaite Ranking study. Bruner applied the two-population replicator dynamic to 5,000 approximating games of 5,000 distinct two-agent bargaining problems. In Bruner's study each of the 5,000 feasible sets had a polygonal structure where the vertex payoffs were sampled at random from a Beta distribution. Bruner found that for the $\boldsymbol{u}_0$-comprehensive feasible set problems of his study the two-population replicator dynamic tended to converge most often to a convention near the Nash solution, while for the noncomprehensive set problems of his study this dynamic tended to converge most often to a convention near the Kalai–Smorodinsky solution (Bruner 2021). Bruner's study powerfully illustrates the importance of $\boldsymbol{u}_0$-comprehensiveness. From Bruner's analysis one might conclude that inductive learning tends to lead towards agents regarding a Nash solution convention as salient in a $\boldsymbol{u}_0$-comprehensive feasible set bargaining problem, and a Kalai–Smorodinsky solution convention as salient in a noncomprehensive feasible set bargaining problem.

Why do bargaining conventions emerge most frequently near the Kalai–Smorodinsky solution for the noncomprehensive feasible set games of Bruner's study and most frequently near the Egalitarian solution for the games of the Braithwaite Ranking study? A complete answer to this question will depend upon future analysis of these sorts of Nash demand games, including perhaps even more systematic computer experiments. But Bruner and I

[58] These percentages are not entirely exclusive since for some of the 1,000 bargaining problems two or more of the Nash, Kalai–Smorodinsky, and Egalitarian solutions either coincide or are very close to each other.

agree that what might look like opposing results in our studies may stem partly from differences in our constructions of the demand games and the conditions we set upon their payoff structures.[59] In the Braithwaite Ranking study, every demand game is built from a basis game where the agents' strict Nash equilibrium payoffs are sufficiently misaligned that each agent prefers some suboptimal outcomes where they receive some positive share over surrendering all of the good to the other. And in the games of this study the outcome where both claim all of the good defines a nonagreement point definitely more favorable to one of the two agents. I analyze games based upon feasible sets with such asymmetries because these tend to produce mutually distinct Nash, Kalai–Smorodinsky, and Egalitarian solutions in the corresponding bargaining problems. Bruner analyzes a more general class of demand games than that of the Braithwaite Ranking study, the former including games constructed from feasible sets where payoffs are not so asymmetric as those of the latter. And in Bruner's study, each demand game is constructed directly from the feasible set, while in the Braithwaite Ranking study each demand game is constructed from a basis game. The direct construction method reflects Nash's idea of allowing the individual payoffs in the feasible set to be the pure strategies of a demand game discussed in Section 3. The basis game construction method reflects Braithwaite's idea of building the feasible set from an extension of a conflictual coordination game that summarizes conflicting preferences over a quantity of a good, also discussed in Section 3. Given these differences, the results of Bruner's and the Braithwaite Ranking study might not be at odds after all. Perhaps these and future related studies may confirm a more general conclusion: Inductive learning interacts with both the structure of the feasible set and the construction of the corresponding demand game to make certain solutions of bargaining problems more salient.

In analyses of bargaining problems such as Skyrms' study, Bruner's study, and the Braithwaite Ranking study discussed above, inductive learning tends to produce bargaining conventions near some solution, like the Kalai–Smorodinsky or the Egalitarian solution, that is in some sense intuitively fair. But such "nice" results presuppose that any member of a population adopts the same strategy for interacting in a demand game with any counterpart member of a population, effectively treating all counterparts equally. What if population members can tailor their demand game strategies according to some observable traits that can vary within populations? In a pioneering study, Robert Axtell, Joshua Epstein, and H. Peyton Young explored inductive learning in a simple

[59] Bruner and I discussed the apparently differing results of these two studies in conversation and private correspondence.

Nash demand game similar to the Figure 29 game among the members of a population having different "tags." These tags had no inherent social significance, but population members could observe counterparts' tags and could condition their claims in the demand game according to them. Axtell, Epstein, and Young showed that their inductive learning model produced outcomes that were both inefficient and had highly asymmetric payoffs across tag groups. As they observed, simply allowing claims conditioned on tags resulted in the emergence of discriminatory norms where one's tag had become socially salient (Axtell, Epstein, and Young 2001). In more recent studies, Bruner (2017), Cailin O'Connor (2019), and Sahar Heydari Fard (2022) have shown that for a variety of learning models and models of assorted interactions applied to Nash demand games, when inductive learners can condition their strategies on initially irrelevant observable traits, discriminatory norms that favor those having some traits at the expense of those having other traits can emerge spontaneously and in many cases tend to emerge most often. Bruner, O'Connor, and Heydari Fard all argue that their findings help explain how at least some of the discriminatory norms based on traits such as gender and ethnicity actually followed in human cultures could have emerged without any deliberate attempts to create them. The results of the research on trait-conditioned strategies in demand games are plainly dramatically different from those of the research on strategies that are effectively blind with respect to varying population traits. All of the specific computer studies discussed in this section illustrate how the results of inductive learning are sensitive to the background conditions that can contribute to the constructions of demand games and to the learning processes themselves.

Studies like those discussed above where one applies a learning process like a replicator dynamic to a game that approximates a bargaining problem examine which bargaining conventions are likely to emerge over the relatively short run. Another and more well-developed research program examines the stability of bargaining conventions over the very long run. This is part of a more general program that explores how certain equilibria of games can be the most stable endpoints of learning processes that are disrupted by random "shocks" or *mutations*.[60] One can interpret such mutations as agents choosing certain strategies different from those they would choose according to the learning process, either by mistake or because they are experimenting. A limit point is *stochastically stable* when the associated learning process approaches and *remains near* this point with positive probability over the very long run in the

[60] Some of the pioneering works in this program are Foster and Young 1990; Kandori, Mailath, and Rob 1993; Young 1993a; and Young 1998.

face of a limited and constant bombardment of mutations (Foster and Young 1990). Stochastic stability is a stronger condition than evolutionary stability. Informally, stochastically stable points are the hardest points to dislodge once the learning process reaches one of these points, in that it takes more simultaneous mutations to "overthrow" a stochastically stable point when this point is incumbent than any point that is not stochastically stable. Several well-known solutions of the bargaining problem are stochastically stable for associated approximating games and different learning dynamics. For example, H. Peyton Young applies an *adaptive learning* dynamic to demand games that approximate Nash bargaining problems. Such adaptive learners update their conjectures regarding counterparts' strategies according to limited samples of the results of past interactions, and then all follow their best responses given their updated conjectures except for those who follow mutant strategies. Young shows that for a two-agent demand game that approximates the demand game of a bargaining problem, the stochastically stable point of adaptive learning is an α-weighted Nash solution (1993b; 1998, 118–124). Young also shows that for a two-agent *contract game* where the two agents must coordinate upon some bargaining convention in order to receive any positive payoffs, the stochastically stable point of adaptive learning is the Kalai–Smorodinsky solution (1998, 131–142). Finally, Young shows that when the agents' encounters in an approximating demand game can be mixed across payoff types, so that much of the time an agent meets a counterpart of a different payoff type but sometimes meets a counterpart of their own payoff type, the stochastically stable state is the convention of simple equal division of the goods at stake (1993b; 1998, 129–130). In a recent study, Sung-Ha Hwang, Wooyoung Lim, Philip Neary, and Jonathan Newton showed that if agents update according to a different *logit choice* rule where there is some intentional bias in the mutations, as if the agents tend to experiment according to certain patterns, then in a two-agent contract game approximating that of a bargaining problem the stochastically stable point is the Egalitarian solution (Hwang et al. 2018). Studies such as these show that long-run stability of bargaining solutions can be sensitive to the payoff structures of the approximating games, the specifics of the learning processes, and the ways in which agents might experiment or make mistakes.

The stability analysis program is of somewhat limited practical value because in most realistic settings, a learning rule like Young's adaptive learning that is bombarded by mutations is likely to require far more time to reach a stochastically stable limit than the expected lifespan of the community of adaptive learners.[61] Indeed, it may be hard if not impossible to test the stochastic stability of a solution to a Nash bargaining problem in the laboratory or in the field. Nevertheless, the long-run stability of certain bargaining conventions such as those characterizing a rule of equal division may also help to explain why the members of many different communities might regard these rules as having a certain salience. As Young puts it, "equal division may be a focal point *because* it is stable, not the other way around" (1998, 130).

7 The Nash Bargaining Problem as a Tool for Analyzing the Social Contract

In the earlier sections of this Element I focused the discussion on the formal structure of Nash's bargaining problem and approaches to identifying solutions of this problem. In this concluding section I will discuss some aspects of the main philosophical application of the bargaining problem, namely the analysis of the social contract. The title of this section is obviously inspired by the title of Richard Braithwaite's 1954 Cambridge lecture: *Game Theory as a Tool for the Moral Philosopher*. In *Evolution of the Social Contract*, Brian Skyrms identifies two broad philosophical approaches for analyzing the social contract, understood in its most generic sense as a body of rules that can regulate interactions among the members of a community. One approach considers which social contract rational agents placed in certain circumstances would choose on behalf of some community. Another approach considers how existing social contracts that regulate actual communities could have emerged ((1996) 2014, xiii). These two rational choice and evolutionary social contract traditions are reflected in the rational choice and conventionalist approaches to finding solutions to Nash's bargaining problem.

Contemporary philosophers and social scientists apply the bargaining problem to contractarian problems ranging from justifying the state to building systems of morality, including especially principles of justice. Here I briefly summarize but a few of these applications. Braithwaite applied the bargaining problem at a micro-level, to develop a single rule for fair division of a good between two claimants. Several philosophers and social scientists following in Braithwaite's footsteps have applied the bargaining problem or its effective equivalent at a more macro-level. In his neo-Hobbesian hypothetical

[61] I discuss this further in Vanderschraaf 2018. Robert Sugden raises the same objection to models of equilibrium selection based on the adaptive learning dynamic in Sugden ((1986) 2004, 203–204.

The adaptive learning dynamic reaches its most persistent long-term limits so slowly because this dynamic sets severe limits on the information of the history of interaction available to agents and incorporates "noise" in the form of independent random errors. Young (1998), 118–143) gives a fine summary of the mechanics of this dynamic applied to bargaining problems.

contractarian account of founding a satisfactory state, Gregory Kavka explicitly names the State of Nature the *no-agreement point* for contractors who fail in the end to reach their *target* of establishing a permanent commonwealth (1986, 189).[62] David Gauthier (1986) and Ken Binmore (1994, 1998) use the bargaining problem to identify general principles of distributive justice for a community having any number of members. Gauthier argues that rational contractors would choose the Maximin Proportionate Gain rule as the principle of distributive justice for their social contract (1986, 129–156; 2013, 609–614).[63] Binmore defends an Egalitarian rule principle of distributive justice using a bargaining problem where claimants have unlimited but constrained renegotiation opportunities, so that any claimant may at any time reselect a solution on behalf of all but without knowing their own position in the bargaining problem. Binmore also argues that over time the social indices that calibrate the utilities of the chosen Egalitarian social contract evolve so that the chosen Egalitarian and the Nash solutions coincide, and then these coincident solutions become the nonagreement point for the next choice of a social contract (1998, 425–447). Ryan Muldoon and Michael Moehler use the bargaining problem to analyze how the members of societies whose diverse moral commitments could if left unchecked produce conflict can find rules for living together peaceably. In Muldoon's and Moehler's contractarian theories, representatives of different subgroups enter into a bargaining problem on behalf of their subgroups. Muldoon and Moehler both argue that the representatives in their bargaining problems resolve the conflicting interests of their subgroups by having them follow some version of the Nash bargaining solution (Muldoon 2016, 68–87; Moehler 2010; Moehler 2018, 84–88).[64] Sugden ((1986) 2004, 69–76, 197–201), Skyrms ((1996) 2014, 1–22), and I (2019, 159–175, 304–319) use the bargaining problem to argue how certain conventions could emerge that their followers come to regard as principles of justice. A leading idea of our analyses is that bargaining conventions that characterize norms widely regarded as norms of fairness are most likely to evolve and remain stable over time.

62 Kavka does not go on to use Nash's mathematical framework of a nonagreement point and feasible set, or any proposed solution concepts for the Nash bargaining problem. Kavka keeps his discussion of founding a state at a conceptual level.

63 Some have taken Gauthier in his 2013 essay to be abandoning bargaining theory altogether while retaining the Maximin Proportionate Gain principle. While I think Gauthier is not too clear on this question, I find it more natural to read him in this later essay as embracing axiomatic analysis rather than strategic bargaining theory.

64 While Muldoon believes the Nash solution summarizes in a rough manner the social contract contractors will choose, he also makes it clear that he does not claim the Nash solution has a great deal of normative force in his theory. He is explicit that other solution concepts might be compatible with his contractarian theory (2016, 79–80).

The bargaining problem is of course not the only device contemporary philosophers use for developing a social contract. John Harsanyi (1953, 1955, 1977) and John Rawls (1963, 1971) established a new tradition of identifying elements of a social contract using a veil of ignorance. More specifically, Harsanyi and Rawls argued that one can identify principles of distributive justice as those principles rational parties would choose while they are constrained by such a veil of ignorance.[65] Tim Scanlon (1982, 1998) proposes that, rather than setting rational agents behind a veil of ignorance, one could constrain the choices such agents would make. In particular Scanlon argues that a satisfactory social contract would include only requirements such rational agents know no member of the relevant community could reasonably reject. The bargaining problem offers an alternative to other approaches to developing the social contract such as the veil of ignorance and the constrained motivation approaches,[66] and this alternative has its advantages. A bargaining problem model makes explicit the baseline against which alternative social contracts are to be evaluated. One way to interpret a bargaining problem is to regard the nonagreement point as characterizing how parties fare at a State of Nature, and the points of the feasible set as characterizing how these parties fare at alternative social contracts they might enter into with respect to this State of Nature. The agents who use the bargaining problem as their tool for choosing a social contract can make their choice with direct reference to the baseline their nonagreement point defines, enabling them to gauge in some sense how much they can all benefit from alternative new social contracts relative to no new social contract. Moreover, since the bargaining problem is defined entirely in terms of utility space, a bargaining problem model provides a framework for directly and systematically comparing the relative merits of alternative social contracts in terms of utility.

In Section 5 I concluded that rational choice analysis is most valuable for clarifying which formal properties one considers essential for a solution to a given bargaining problem. In selecting a solution concept for a bargaining problem in a given context, a rational choice theorist must decide which formal desiderata they believe are essential and which can be "sacrificed." I believe the

[65] As is well known, Harsanyi and Rawls derive different and incompatible principles of distributive justice in their respective rational choice social contract theories, partly because they employ somewhat different veils of ignorance. Moehler gives an illuminating recent reappraisal of the Harsanyi–Rawls dispute in *Minimal Morality* (2018, 67–73).

[66] In fact, Binmore's bargaining problem–based social contract theory employs a veil of ignorance, since any claimant who chooses to reselect the bargaining problem solution at a given time must do so without knowing their identity in the bargaining problem. Harsanyi's and Rawls' veil of ignorance theories differ from Binmore's approach in that in Harsanyi's and Rawls' theories, parties may choose exactly once while constrained by a veil of ignorance and they choose one of a set of alternative social welfare functions rather than a solution of a bargaining problem.

rational choice approach offers an analogous moral for social philosophers. Some have thought that philosophers like Braithwaite and Gauthier who use the bargaining problem in their social contract theories aspire to derive a social contract solely from principles of rational choice. I think this is a misunderstanding, one similar to the mistaken view that Rawls and Harsanyi are trying to create principles of justice solely from principles of rational choice.[67] I think rational choice bargaining theory does its real work for social philosophers by clarifying the most fundamental moral commitments one incorporates into a given social contract. Indeed, axiomatic bargaining theorists themselves observe that a number of proposed axioms have some underlying moral motivation. As noted before, Pareto Optimality reflects the idea that one should avoid unnecessary waste. Strong Monotonicity reflects an evidently minimal prerequisite of a mutually advantageous social contract: If the cooperative surplus grows, then everyone should fare at least as well as they fared before. Population Monotonicity has its intuitive appeal if one interprets the feasible set as generated by a fixed quantity of goods where the number of claimants might vary, so that no one is entitled to more than their original share of this fixed "pie" simply because now more claimants are present. And one can give similar moral motivations for many of the other well-known axioms. Philosophers can incorporate fundamental moral commitments into their bargaining problem social contract theories by their choices of particular axioms. Another way they incorporate such moral commitments is by setting specific constraints upon the nonagreement point.

Gauthier's and Moehler's social contract theories are fine illustrations of how philosophers make clear their core moral commitments using the bargaining problem as their foil. Gauthier adopts the Maximin Proportionate Gain solution and requires his nonagreement point to satisfy his *Lockean Proviso*, which reflects a Lockean State of Nature where agents have free and equal access to resources and also have some personal rights and some property rights. As Gauthier forthrightly acknowledges, the Maximin Proportionate Gain solution is neither Contraction Consistent nor Strongly Monotone, though it is Symmetric, Pareto Optimal, and Scale Invariant (1985, 36–37). Gauthier

[67] Gauthier perhaps unintentionally encourages this mistaken understanding of his rational choice social contract since early in *Morals by Agreement* he declares that a central part of his project in this work is to generate moral principles from nonmoral premises of rational choice (1986, 4–6). However, Gauthier incorporates some initial moral commitments into his social contract theory, most obviously when he requires that the nonagreement point of the bargaining problem that parties use to identify this social contract satisfies his Lockean Proviso (1986, 190–232).

Rawls, like Gauthier, may have unintentionally encouraged a similar misinterpretation of his theory since in at least one place Rawls says that justice is a part of the theory of rational choice (1971, 16).

willingly sacrifices any Contraction Consistency requirement because he believes his contractors are concerned with maximizing their own individual gains and not with maximizing any joint gain such as that of the Nash product. And Gauthier also willingly sacrifices Strong Monotonicity in order to ensure Pareto Optimality together with Symmetry and Scale Invariance, which he regards as essential properties of the outcome of morals chosen by rational contractors. Gauthier designs his bargaining problem social contract this way in the belief that his specific requirements provide the best basis for developing a full system of morals understood as constraints on individual conduct that when generally followed are to the advantage of all.

For the central part of his social contract theory, Moehler adopts a refinement of the Nash bargaining solution, the *Stabilized Nash solution*, that makes a *minimum standard of living* point $\boldsymbol{m}_0$ the point of reference for maximizing the Nash product while retaining an original "free for all" point $\boldsymbol{u}_0$ as nonagreement point (2010; 2018, 84–88). The underlying intuition of Moehler's Stabilized Nash solution is that if at any proposed agreement the claimants do not all fare at least as well as each would fare at $\boldsymbol{m}_0$, then those who do not achieve even their minimum standard of living are prepared to bring all to $\boldsymbol{u}_0$, a nonagreement point reflecting a Hobbesian State of Nature war. Moehler adopts this specific solution concept because he aims to incorporate as little in the way of background moral assumptions as possible. He argues his refinement of the Nash solution concept is best suited for this purpose. Moehler also clearly acknowledges that consequently this part of his social contract theory is more limited in scope than a theory like Gauthier's, and in particular is best suited for setting only minimal standards of peaceful interaction between parties who may initially agree on no moral matters at all (2018, 23, 158–165). Moehler's Stabilized Nash solution concept is neither Strongly Monotone nor Population Monotone, so his social contract is not renegotiation-proof with respect to changes in the feasible set. But the Stabilized Nash solution concept is Scale Invariant, so for Moehler's contractarian theory where parties are not presumed to agree upon questions of value, this concept has the advantage of not assuming any common standards by which the parties could interpersonally compare utilities. Moehler accepts the price that the terms of cooperation may sometimes need to be rewritten when background conditions change, so that he can produce a social contract theory that can serve parties who deeply disagree upon questions of values.

In a response to Moehler's work, I argue that contractors, in a slightly different bargaining situation or with a slightly different understanding of the bargaining problem described by Moehler, might be willing to consider a wider range of social contracts than Moehler's Stabilized Nash solution contract.

More specifically, I argue that contractors might possibly select either a social contract where in their bargaining problem they follow the Kalai–Smorodinsky solution or a ***m_0**-stabilized Asymmetric Nash solution* with a weighted Nash product similar to that of the **α**-weighted Nash solution. The **m_0**-stabilized Asymmetric Nash solution is Reapplication Stable, though neither Strongly Monotone nor Population Monotone. The Kalai–Smorodinsky solution is Population Monotone, though not Reapplication Stable. I maintain that contractors might prefer a **m_0**-stabilized Asymmetric Nash contract if on the one hand they can easily try to renegotiate their shares across subgroups at any point in time, and on the other hand changes in the feasible set occur only relatively infrequently. For they might then regard the possibility that some could lose out solely on account of changes in the feasible set as an acceptable price for having a social contract that is renegotiation-proof across subgroups. Conversely, I maintain that contractors might prefer a Kalai–Smorodinsky social contract if certain changes in the feasible set occur relatively often and in fact they cannot so easily renegotiate the terms of this contract across subgroups. For then they might be willing to accept the possible instability of this social contract when negotiations across subgroups do happen to occur, in order to enjoy the relatively good stability properties of the Kalai–Smorodinsky solution with respect to changes in the feasible set (Vanderschraaf 2020). As examples such as these show, rational choice bargaining theory can serve as scaffolding for quite a range of social contract theories.

Rational choice contractarian theories are characteristically normative, so it is not surprising that contractarians in this tradition who use the bargaining problem use rational choice bargaining theory, which is motivated by the question, "What outcome of a bargaining problem would rational agents select?" Evolutionary contractarian theories are predominantly explanatory. Evolutionary contractarians tend to rely upon evolutionary models of bargaining problem encounters. Such models are motivated by the question, "Which bargaining conventions are likely to emerge and remain in force over time?" The focal point and inductive learning examples discussed in earlier sections illustrate a general conclusion: Inductive learning and focal point reasoning complement each other and can generate a variety of different conventions for the same bargaining problem. At the same time, these examples also show that not all bargaining conventions must be equally likely to emerge. They illustrate how parties in a Hobbesian State of Nature situation where they have no antecedent obligations to treat each other with restraint can learn to follow certain principles of fair division. In this sense, rules for "playing fair" can be the product of inductive learning.

A related and admittedly more tentative general conclusion is that conventionalist analysis may shed light upon how attempts at interpersonal utility comparisons are related to certain important general principles of justice. Philosophers and social scientists differ sharply in their views regarding the very idea of interpersonally comparable utilities. Some doubt that a sound account of interpersonally comparable utilities is possible. Nash for one rejected the idea. Nash formulated the bargaining problem in terms of von Neumann–Morgenstern utilities, and noted that his Scale Invariance axiom precludes the possibility of interpersonally comparable utilities (1953, 137). Yet others, like John Harsanyi, argue that interpersonally comparable utilities do make sense and that they are necessary for moral theorizing. Harsanyi goes so far as to claim that one must at least attempt interpersonal utility comparisons if one is to make moral value judgments rationally (1977, 81–82).[68] A number of scholars, including Harsanyi himself (1955, 316–321; 1977, 48–83), have tried to develop rigorous logical foundations for interpersonal utility comparisons. But all such attempts incorporate fundamental assumptions not open to empirical proof, and skepticism regarding the possibility of a sound account of interpersonally comparable utilities persists.

Still, I think Harsanyi is plainly right to maintain that people routinely try to make interpersonal utility comparisons in their everyday affairs (1955, 316–317). Perhaps people ordinarily try to make such comparisons only in a rough manner, with no conscious use of formal utility theory. But they do try. Luce and Raiffa and William Thomson point out that in many actual bargaining situations the parties involved make informal references to interpersonal utility comparisons. I agree with Luce and Raiffa and Thomson that in such cases it may be appropriate to apply a solution concept that employs interpersonal utility comparisons (Luce and Raiffa 1957, 131–132; Thomson 1994, 1247; Thomson 2010, xviii). I think conventionalist analysis of some bargaining problems may shed light upon how such actual attempts at comparing utilities across different individuals contribute to the origins of certain general and well-known principles of justice. For example, Aristotle claims everyone agrees upon a *proportionality principle* of distributive justice, according to which the ratios of the values of shares received and of the recipients' worth are equal (2014, 1131a20–1131b22; 2017, 1280a8–1280a30, 1282b18–1282b22). Aristotle gives no defense of this formal principle or any evidence that this principle is indeed universally accepted, and appears to think that no such defense or evidence is necessary. In fact, Aristotle's proportionality principle

[68] However, it is important to note that Harsanyi doubts that interpersonal utility comparisons should play a direct role in the solution of a bargaining problem (1977, 192–195). Unlike Raiffa and Braithwaite, Harsanyi does not consider the bargaining problem a proper tool for analyzing principles of fair division.

is formally equivalent to the rule defining the Egalitarian solution of the Nash bargaining problem (Binmore 1998, 397–399; Vanderschraaf 2018, 259–260; Vanderschraaf 2019, 187–188).[69] This solution concept presupposes interpersonally comparable utilities that are calibrated according to some common scale. Only given this presupposition can it make sense to refer to points where all achieve equal gains with respect to the nonagreement point. When he introduced the Egalitarian solution, Raiffa converted each agent's payoffs to a 0–1 scale. One way to interpret the grounds for this particular conversion is to assume that each agent receives the same positive benefit at the outcome most favorable to them and no positive benefit at the outcome least favorable to them. The computer simulation studies discussed in this Element are based upon bargaining problems with such a 0–1 scale. The results of these studies show that when the agents' interests in a two-agent bargaining problem are sufficiently misaligned and their payoffs are scaled this way, the Egalitarian solution has a certain "pull" in the learning process. For this class of bargaining problems learning can contribute to making the Egalitarian solution a focal bargaining convention. This in turn may help to explain why the Aristotelian proportionality principle would be widely accepted.

To close this section I will state what I take to be the two most important open problems in the bargaining problem social contract research program. On the rational choice side, philosophers need to explore more extensively how certain background social conditions map to some solution of a bargaining problem. I believe that no one bargaining problem solution concept should serve to characterize all social contracts. Gauthier's Maximin Proportionate Gain social contract, Binmore's Egalitarian social contract, and Moehler's Stabilized Nash solution social contract all illustrate how different bargaining problem social contracts can serve different communities with differing needs. But I think philosophers can and should move beyond such specific cases and towards a more general account of the conditions that characterize particular bargaining problem social contracts.[70]

---

69 My arguments for this equivalence are based upon David Keyt's fine reconstruction of Aristotle's account of distributive justice (Keyt 1991).

70 John Roemer has taken an interesting, though largely overlooked, step in this direction. Roemer argues that by formulating the bargaining problem solely in terms of utility space, which I have claimed gives the bargaining problem its great flexibility in applications, Nash has excluded information crucially relevant to distributive justice. Roemer proposes to restore this information by embedding the Nash bargaining problem in an *economic environment* that formally models the goods and the lotteries that generate the feasible set. This embedding is similar to the idea of generating a feasible set from a basis game, but is meant to give a fully general formal framework for modeling distributive justice problems (1986; 1996, 78–91, 96–103). With his economic environment framework Roemer derives an Egalitarian solution–style social contract for a specific domain of problems (1996, 103–119). But in the end he argues that even this richer formal framework is unlikely to capture all the relevant information in all distributive justice problems (1996, 123–125).

On the conventionalist side, bargaining theorists need to explore in depth which if any solutions are more likely to emerge as bargaining conventions in problems having more than two claimants. Most of the now vast body of laboratory experiments, field studies, and computational studies of bargaining problems focus on a two-agent format. While the two-agent case is certainly very important, the real test of conventionalist analysis may lie in the analysis of bargaining problems with three or more claimants. Researchers should explore which, if any, bargaining conventions in *n*-agent bargaining problems where $n \geq 3$ are most stable with respect to adding or removing agents. Moreover, when agents learn inductively according to their experiences of past interactions, when $n \geq 3$ they may consider the possibility that the claims of their counterparts are *correlated.* For simple three-agent conflictual coordination games, the possibility of such *correlated learning* can have a profound effect upon the agents' prospects for reaching any of the available optimal equilibria (Vanderschraaf and Richards 1997). I conjecture that empirical and computer studies of $n \geq 3$ bargaining problems that can incorporate any or all of these possibilities may profoundly change our understanding of how certain bargaining conventions tend to become salient.

Appendix: Braithwaite Ranking Simulations

The Braithwaite Ranking computer simulation study discussed in Section 6 is based upon 1,000 distinct games, each of which approximates the Nash demand game of one of 1,000 distinct bargaining problems. The constructions of these 1,000 approximating games and the replicator dynamic simulations were performed using MatLab 9.[1] Each of these 1,000 bargaining problems is constructed from a basis game having a payoff structure summarized in Figure A1. Each Figure A1 basis game generates a bargaining problem with the 0–1 payoff scaling that Luce and Raiffa apply to the Braithwaite basis game and the resulting bargaining problem and game (1957, 146–147). The requirement that $\frac{1}{2} \geq \alpha_i > \beta_i > 0, i \in \{1,2\}$ ensures that each basis game has a payoff structure reflecting ordinal preferences over the pure strategy outcomes equivalent to those of the Figure 11 Braithwaite basis game. When the Figure A1 game is interpreted as a division problem as in Braithwaite's Cambridge lecture, this requirement ensures that: (i) each of the two claimants gains some satisfaction from the other's receiving some of the good at stake, so that neither has purely nontuistic preferences, and (ii) their preferences are also sufficiently misaligned so that, assuming interpersonally comparable utilities, the convex hull of $(\beta_1, 0)$, $(1, \alpha_2)$, $(\alpha_1, 1)$, and $(0, \beta_2)$, the respective payoff vectors of (M, M), (G, M), (M, G), and (G, G), contains payoff vectors that each claimant strictly prefers over some payoff vectors where both claimants would make equal utility gains from $(0, \beta_2)$.[2] The Figure 11 game is the instance of the Figure A1 game with $\alpha_1 = \frac{1}{2}, \beta_1 = \frac{1}{6}, \alpha_2 = \frac{2}{9}$. Like the Braithwaite basis game, each of these 1,000 Figure A1 basis games is a conflictual coordination game with payoff asymmetries such that (G, M) is the outcome most favorable to Agent 1, (M, G) is the outcome most favorable to Agent 2, all of the Pareto frontier outcomes defined by $\lambda \cdot (M, G) + (1 - \lambda) \cdot (G, M), \lambda \in [0, 1]$ are strict correlated equilibria, and the

[1] The MatLab programs used in this simulation study and the complete set of simulation statistics recorded for these 1,000 approximating games are available from the author upon request.

[2] In Bruner's related simulation study (2021), Bruner sets less restrictive constraints on the set of games he analyzes that have a conflictual coordination structure. So in the conflictual coordination games Bruner analyzes, it is possible for the claimants' interests to be misaligned to a lesser degree than they are in the Braithwaite Ranking study. In private discussion Bruner and I agreed that one interesting consequence of his less restrictive constraints is that in many of the games of his study the Egalitarian solution is the nonagreement point. The more stringent restrictions I set on the payoffs of the Braithwaite Ranking study guarantee that the Egalitarian solution is on the Pareto frontier. Bruner and I agree that this difference may explain in part the differing results of our simulation studies.

		Agent 2	
		M	G
Agent 1	M	$(\beta_1, 0)$	$(\alpha_1, 1)$
	G	$(1, \alpha_2)$	$(0, \beta_2)$

M = modest (claim none), G = greedy (claim all)

$$\tfrac{1}{2} \geq \alpha_i > \beta_i > 0$$

Figure A1 Braithwaite Ranked basis game

(G, G) outcome is for Agent 1 the worst possible outcome but for Agent 2 only the second worst possible outcome. This last condition ensures that the non-agreement point $(0, \beta_2)$ is less unfavorable to Agent 2 than to Agent 1, so that Agent 2 has a threat advantage. For each game, the payoffs for each Agent i are determined by randomly sampling values from a uniform distribution over the unit interval $[0, 1]$ and then reordering and normalizing these sampled values so that they conform to the Figure A1 game payoffs.

For each approximating demand game, the claim precision was $\delta = \frac{1}{50}$. The individual population members in the roles of Agent 1 and Agent 2 each could follow one of fifty-one pure strategies $s_{i,0}, s_{i,1}, \ldots, s_{i,50}$ where Agent i follows $s_{i,m}$ by claiming $m \cdot \delta$ of the good, $m \in \{0, 1, \ldots, 50\}$. The resulting strategic form game had $51^2 = 2601$ pure strategy profiles, and approximated the completely continuous bargaining problem quite well in that each agent could claim any even whole number percentage of the good. This game had fifty-one Nash equilibria in pure strategies of the form $(s_{1,m}, s_{2,(50-m)})$ where $m \in \{0, 1, \ldots, 50\}$, so that Agent 1 follows $s_{1,m}$ and claims $m \cdot \delta$ and Agent 2 follows $s_{2,(50-m)}$ and claims $(50 - m) \cdot \delta$. For $m \in \{1, \ldots, 49\}$, at each of these $(s_{1,m}, s_{2,(50-m)})$ equilibria both claimants gain some positive share of the good at stake. The remaining two pure strategy equilibria $(s_{1,50}, s_{2,0})$ and $(s_{1,0}, s_{2,50})$ respectively characterize the Nash equilibria of the basis game (G, M) where Agent 1 is greedy and Agent 2 is modest, and (M, G) where Agent 1 is modest and Agent 2 is greedy. All of these $(s_{1,m}, s_{2,(50-m)})$ equilibria are strict and correspond to efficient divisions of the good at stake. These equilibria characterize fifty-one different basic bargaining conventions that are possible for each of these approximating games.

To estimate the relative attracting power of the different bargaining conventions for each of the 1,000 approximating games, for each of these games I simulated 1,000 orbits each for the two-population replicator dynamic, where the members of one population had Agent 1's payoffs and the members of the other population had Agent 2's payoffs. Each of these 1,000 orbits was set at an initial pure strategy distribution that was chosen randomly in the distribution simplex. In all simulations taken across the 1,000 approximating games the two-population replicator dynamic converged to some bargaining convention equilibrium. Each bargaining convention characterized by an $(s_{1,m}, s_{2(50-m)})$ equilibrium had a basin of attraction $\Delta_{(s_{1,m}, s_{2,(50-m)})}$ of the two-population replicator dynamic. I estimated the relative sizes of these basins as the relative frequencies of orbits that converged to given $(s_{1,m}, s_{2,(50-m)})$ convention equilibria. For each of the 1,000 bargaining problems generated by the 1,000 different Figure A1 games I computed the Nash, the Kalai–Smorodinsky, and the Egalitarian solutions. As reported in Section 6, for 39.7 percent of the cases (397 of the 1,000 games) the bargaining convention with the largest estimated basin of attraction was within $\delta = \frac{1}{50}$ of the Egalitarian solution, for 15.9 percent of the cases (159 of the 1,000 games) the bargaining convention with the largest estimated basin of attraction was within δ of the Nash solution, and for 14.1 percent of the cases (141 of the 1,000 games) the bargaining convention with the largest estimated basin of attraction was within δ of the Kalai–Smorodinsky solution. For 45.7 percent of the cases, the bargaining convention with the largest estimated basin of attraction was not within δ of any of the Nash, Kalai–Smorodinsky, or Egalitarian solutions. As noted in Section 6 there is some overlap in these totals because in some cases the Nash, Kalai–Smorodinsky, or Egalitarian solutions may coincide or be very close to each other, so that more than one of these three solutions may be within δ of the largest estimated basin of attraction.[3] So overall, for the two-population replicator dynamic in these simulations the Egalitarian solution exhibits greater attracting power than either the Nash or the Kalai–Smorodinsky solutions, but in these simulations bargaining conventions also frequently emerge that are not within δ of any of these three well-known solutions. From the conventionalist perspective this is not surprising. For from this perspective, for a given bargaining problem a population might converge into one of many different bargaining conventions, though certain distinguished conventions may emerge more frequently than others.

[3] See note 58.

References

Alexander, Jason McKenzie, and Skyrms, Brian. 1999. "Bargaining with neighbors: is justice contagious?" *Journal of Philosophy* 96: 588–598.

Aristotle. 2014. *Nicomachean Ethics*. Trans. C. D. C. Reeve. Indianapolis, IN: Hackett Publishing Company.

Aristotle. 2017. *Politics*. Trans. C. D. C. Reeve. Indianapolis, IN: Hackett Publishing Company.

Arrow, Kenneth J. (1951) 2012. *Social Choice and Individual Values*, 3rd ed. New Haven, CT: Yale University Press.

Axtell, Robert L., Epstein, Joshua M., and Young, H. Peyton. 2001. "The emergence of classes in a multi-agent bargaining problem." In Steven N. Durlauf and H. Peyton Young, eds., *Social Dynamics*. Washington, DC: Brookings Institution Press, 191–211.

Bardsley, Nicholas, Mehta, Judith, Starmer, Chris, and Sugden, Robert. 2010. "Explaining focal points: cognitive hierarchy theory versus team reasoning." *Economic Journal* 120: 40–79.

Barry, Brian. 1989. *Theories of Justice*. Berkeley: University of California Press.

Binmore, Ken. 1987. "Perfect equilibria in bargaining models." In Ken Binmore and Partha Dasgupta, eds., *The Economics of Bargaining*. Oxford: Basil Blackwell, 61–76.

Binmore, Ken. 1994. *Game Theory and the Social Contract*, vol. 1: *Playing Fair*. Cambridge, MA: MIT Press.

Binmore, Ken. 1998. *Game Theory and the Social Contract*, vol. 2: *Just Playing*. Cambridge, MA: MIT Press.

Binmore, Ken, Rubinstein, Ariel, and Wolinsky, Asher. 1986. "The Nash bargaining solution in economic modeling." *RAND Journal of Economics* 117: 176–188.

Bossert, Walter, and Tan, Guofu. 1995. "An arbitration game and the egalitarian solution." *Social Choice and Welfare* 12: 29–41.

Braithwaite, Richard. (1955) 1994. *Theory of Games as a Tool for the Moral Philosopher*. Bristol: Thoemmes Press.

Bruner, Justin. 2017. "Minority (dis)advantage in population games." *Synthese* 196: 413–427.

Bruner, Justin. 2021. "Nash, bargaining and evolution." *Philosophy of Science* 88: 1185–1198.

Camerer, Colin. 2003. *Behavioral Game Theory: Experiments in Strategic Interaction*. Princeton, NJ: Princeton University Press.

Foster, Dean, and Young, H. Peyton. 1990. "Stochastic evolutionary game dynamics." *Theoretical Population Biology* 38: 219–232.

Gaertner, Wulf, and Klemisch-Ahlert, Marlies. 1992. *Social Choice and Bargaining Perspectives on Distributive Justice*. Berlin: Springer Verlag.

Gauthier, David. 1985. "Bargaining and justice." *Social Philosophy & Policy* 2: 29–47.

Gauthier, David. 1986. *Morals by Agreement*. Oxford: Clarendon Press.

Gauthier, David. 2013. "Twenty-five on." *Ethics* 124: 601–624.

Harsanyi, John. 1953. "Cardinal utility in welfare economics and the theory of risk taking." *Journal of Political Economy* 61: 343–345.

Harsanyi, John. 1955. "Cardinal welfare, individualistic ethics, and interpersonal comparisons of utility." *Journal of Political Economy* 63: 309–321.

Harsanyi, John. 1977. *Rational Behavior and Bargaining Equilibrium in Games and Social Situations*. Cambridge: Cambridge University Press.

Harsanyi, John, and Selten, Reinhard. 1972. "A generalized Nash solution for two-person bargaining games with incomplete information." *Management Science* 18: 80–106.

Heydari Fard, S. 2022. "Strategic injustice, dynamic network formation, and social movements." *Synthese* 200: 392.

Hobbes, Thomas. (1651) 1994. *Leviathan*. Ed. Edwin Curley. Indianapolis, IN: Hackett Publishing Company.

Hume, David. (1740) 2000. *A Treatise of Human Nature*. Ed. David Fate Norton and Mary J. Norton. Oxford: Oxford University Press.

Hume, David. (1751) 1998. *An Enquiry Concerning the Principles of Morals: A Critical Edition*. Ed. Tom Beauchamp. Oxford: Clarendon Press.

Hwang, Sung-Ha, Lim, Wooyoung, Neary, Philip, and Newton, Jonathan. 2018. "Conventional contracts, intentional behavior and logit choice: equality without symmetry." *Games and Economic Behavior* 110: 273–294.

Imai, Haruo. 1983. "Individual monotonicity and lexicographic maximin solution." *Econometrica* 51: 389–401.

Kalai, Ehud. 1977. "Proportional solutions to bargaining situations: interpersonal utility comparisons." *Econometrica* 44: 1623–1630.

Kalai, Ehud, and Smorodinsky, Meir. 1975. "Other solutions to Nash's bargaining problem." *Econometrica* 16: 29–56.

Kandori, Michihiro, Mailath, George J., and Rob, Rafael. 1993. "Learning, mutation, and long run equilibria in games." *Econometrica* 61: 29–56.

Kavka, Gregory. 1986. *Hobbesian Moral and Political Theory*. Princeton, NJ: Princeton University Press.

Keyt, David. 1991. "Aristotle's theory of distributive justice." In David Keyt and Fred D. Miller, Jr., eds., *A Companion to Aristotle's Politics*. Oxford: Basil Blackwell, 238–278.

Krishna, Vijay, and Serrano, Roberto. 1996. "Multilateral bargaining." *Review of Economic Studies* 63: 61–80.

Lewis, David. 1969. *Convention: A Philosophical Study*. Cambridge, MA: Harvard University Press.

Lucas, John R. 1959. "Moralists and gamesmen." *Philosophy* 34: 1–11.

Luce, R. Duncan, and Raiffa, Howard. 1957. *Games and Decisions: Introduction and Critical Survey*. New York: John Wiley and Sons.

Maynard Smith, John. 1982. *Evolution and the Theory of Games*. Cambridge: Cambridge University Press.

Mehta, Judith, Starmer, Chris, and Sugden, Robert. 1992. "An experimental investigation of focal points in coordination and bargaining: some preliminary results." In John Geweke, ed., *Decision Making under Risk and Uncertainty: New Models and Empirical Findings*. Dordrecht: Kluwer Academic Publishers, 211–219.

Mehta, Judith, Starmer, Chris, and Sugden, Robert, 1994a. "The nature of salience: an experimental investigation." *American Economic Review* 84: 658–673.

Mehta, Judith, Starmer, Chris, and Sugden, Robert, 1994b. "Focal points in pure coordination games: an experimental investigation." *Theory and Decision* 36: 163–185.

Moehler, Michael. 2010. "The (stabilized) Nash bargaining solution as a principle of distributive justice." *Utilitas* 22: 447–473.

Moehler, Michael. 2018. *Minimal Morality: A Multilevel Social Contract Theory*. Oxford: Oxford University Press.

Moulin, Herve. 1984. "Implementing the Kalai–Smorodinsky Solution." *Journal of Economic Theory* 33: 32–45.

Muldoon, Ryan. 2016. *Social Contract Theory for a Diverse World*. New York: Routledge Taylor & Francis Group.

Nash, John. 1950. "The bargaining problem." *Econometrica* 18: 155–162.

Nash, John. 1951a. "Non-cooperative games." *Annals of Mathematics* 54: 286–295.

Nash, John. (1951b) 1996. "Appendix: motivation and interpretation." In *Essays on Game Theory*. Cheltenham, UK: Edward Elgar, 32–33.

Nash, John. 1953. "Two-person cooperative games." *Econometrica* 21: 128–140.

O'Connor, Cailin. 2019. *The Origins of Unfairness: Social Categories and Cultural Evolution*. Oxford: Oxford University Press.

Osborne, Martin J., and Rubinstein, Ariel. 1990. *Bargaining and Markets*. San Diego, CA: Academic Press Inc.

Peters, Hans. 1992. *Axiomatic Bargaining Game Theory*. Dordrecht: Springer Science+Business Media.

Raiffa, Howard. 1951. "Arbitration schemes for generalized two-person games." Doctoral dissertation, University of Michigan.

Raiffa, Howard. 1953. "Arbitration schemes for generalized two-person games." In Harold William Kuhn and Albert William Tucker, eds., *Contributions to the Theory of Games*, vol. 2, Annals of Mathematics Studies no. 28. Princeton, NJ: Princeton University Press, 361–387.

Rawls, John. 1958. "Justice as fairness." *Philosophical Review* 67: 164–194.

Rawls, John. 1963. "The sense of justice." *Philosophical Review* 72: 281–305.

Rawls, John. 1971. *A Theory of Justice*. Cambridge, MA: Harvard University Press.

Roemer, John. 1986. "The mismarriage of bargaining theory and distributive justice." *Ethics* 97: 88–110.

Roemer, John. 1996. *Theories of Distributive Justice*. Cambridge, MA: Harvard University Press.

Roth, Alvin. 1977. "Individual rationality and Nash's solution to the bargaining problem." *Mathematics of Operations Research* 2: 64–65.

Roth, Alvin. 1979. *Axiomatic Models of Bargaining*. Berlin: Springer Verlag.

Roth, Alvin. 1985. "Toward a focal-point theory of bargaining." In Alvin E. Roth, ed., *Game Theoretic Models of Bargaining*. Cambridge: Cambridge University Press, 259–268.

Roth, Alvin. 1995. "Bargaining experiments." In John H. Kagel and Alvin E. Roth, eds., *Handbook of Experimental Economics*. Princeton, NJ: Princeton University Press, 253–348.

Roth, Alvin, and Murnighan, Keith. 1982. "The role of information in bargaining: an experimental study." *Econometrica* 50: 1123–1142.

Rubinstein, Ariel. 1982. "Perfect equilibrium in a bargaining model." *Econometrica* 50: 97–109.

Scanlon, T. M. 1982. "Contractualism and utilitarianism." In Amartya Sen and Bernard Williams, eds., *Utilitarianism and Beyond*. Cambridge: Cambridge University Press, 103–128.

Scanlon, T. M. 1998. *What We Owe to Each Other*. Cambridge, MA: Harvard University Press.

Schelling, Thomas. 1960. *The Strategy of Conflict*. Cambridge, MA: Harvard University Press.

Sen, Amartya. (1970) 2017. *Collective Choice and Social Welfare: An Expanded Edition*. Cambridge, MA: Harvard University Press.

Skyrms, Brian. (1996) 2014. *Evolution of the Social Contract*, 2nd ed. Cambridge: Cambridge University Press.

Smith, Adam. (1759) 1982. *The Theory of Moral Sentiments*. Ed. D. D. Raphael and A. L. MacFie. Indianapolis, IN: Liberty Fund.

Ståhl, Ingolf. 1972. *Bargaining Theory*. Stockholm: Economic Research Institute.

Sugden, Robert. (1986) 2004. *The Economics of Rights, Co-operation and Welfare*, 2nd ed. Houndsmills, Basingstoke: Palgrave Macmillan.

Thomson, William. 1994. "Cooperative models of bargaining." In Robert J. Aumann and Sergiu Hart, eds., *Handbook of Game Theory with Economic Applications*, vol. 2. Amsterdam: Elsevier B. V., 1237–1284.

Thomson, William. 2010. "Introduction." In William Thomson, ed., *Bargaining and the Theory of Cooperative Games: John Nash and Beyond*. Cheltenham, UK: Edward Elgar, xv–li.

Thomson, William, and Lensberg, Terje. 1989. *Axiomatic Theory of Bargaining with a Variable Number of Agents*. Cambridge: Cambridge University Press.

Thrasher, John. 2014. "Uniqueness and symmetry in bargaining theories of justice." *Philosophical Studies* 167: 683–699.

Van Damme, Eric, Selten, Reinhard, and Winter, Eyal. 1990. "Alternating bid bargaining with a smallest money unit." *Games and Economic Behavior* 2: 188–201.

Vanderschraaf, Peter. 2018. "Learning bargaining conventions." *Social Philosophy & Policy* 35: 237–263.

Vanderschraaf, Peter. 2019. *Strategic Justice: Convention and Problems of Balancing Divergent Interests*. New York: Oxford University Press.

Vanderschraaf, Peter. 2020. "Stability challenges for Moehler's second-level social contract." *Analytic Philosophy* 61: 70–86.

Vanderschraaf, Peter, and Richards, Diana. 1997. "Joint beliefs in conflictual coordination games." *Theory and Decision* 42: 287–310.

Von Neumann, John, and Morgenstern, Oskar. (1944) 2004. *Theory of Games and Economic Behavior, Sixtieth-Anniversary Edition*. Princeton, NJ: Princeton University Press.

Young, H. Peyton. 1993a. "The evolution of conventions." *Econometrica* 61: 57–84.

Young, H. Peyton. 1993b. "An evolutionary theory of bargaining." *Journal of Economic Theory* 59: 145–168.

Young, H. Peyton. 1994. *Equity in Theory and Practice*. Princeton, NJ: Princeton University Press.

Young, H. Peyton. 1998. *Individual Strategy and Social Structure: An Evolutionary Theory of Institutions*. Princeton, NJ: Princeton University Press.

Acknowledgments

I am grateful to Martin Peterson, the editor of this Element series, for his unflagging aid and support through every stage of the process of preparing this Element. I am also grateful to three anonymous referees for Cambridge University Press, whose excellent comments on an earlier version of this Element led me to substantially rewrite large parts and to significantly improve the overall text. Justin Bruner and William Thomson in conversation and in correspondence gave me invaluable help in refining discussions of some of the finer points of bargaining theory. Finally, I am grateful to my spouse Claudia Vanderschraaf for her patience and steadfast support throughout the lengthy period I worked on this project.

In memory of Gerald Gaus and William Keech

Cambridge Elements

Decision Theory and Philosophy

Martin Peterson
Texas A&M University

Martin Peterson is Professor of Philosophy and Sue and Harry E. Bovay Professor of the History and Ethics of Professional Engineering at Texas A&M University. He is the author of four books and one edited collection, as well as many articles on decision theory, ethics and philosophy of science.

About the Series

This Cambridge Elements series offers an extensive overview of decision theory in its many and varied forms. Distinguished authors provide an up-to-date summary of the results of current research in their fields and give their own take on what they believe are the most significant debates influencing research, drawing original conclusions.

Cambridge Elements

Decision Theory and Philosophy

Elements in the Series

Dutch Book Arguments
Richard Pettigrew

Rational Choice Using Imprecise Probabilities and Utilities
Paul Weirich

Beyond Uncertainty: Reasoning with Unknown Possibilities
Katie Steele and H. Orri Stefánsson

Evidential Decision Theory
Arif Ahmed

Commitment and Resoluteness in Rational Choice
Chrisoula Andreou

Money-Pump Arguments
Johan E. Gustafsson

Bargaining Theory
Peter Vanderschraaf

A full series listing is available at: www.cambridge.org/EDTP